Henry McCormick

Suggestions on teaching Geography

Henry McCormick

Suggestions on teaching Geography

ISBN/EAN: 9783337166700

Printed in Europe, USA, Canada, Australia, Japan

Cover: Foto ©Andreas Hilbeck / pixelio.de

More available books at **www.hansebooks.com**

SUGGESTIONS

ON

TEACHING GEOGRAPHY

BY

HENRY MCCORMICK

Professor of History and Geography in the Illinois Normal
University

Author of Practical Work in Geography

STATE NORMAL SCHOOL,
LOS ANGELES, CAL.

CONTENTS.

INTRODUCTORY.

	PAGE.
CHAPTER I. Why Class Geography Among the Sciences?	7
CHAPTER II. The Content of Geography,	13
CHAPTER III. The Educational Value of Geography,	19
CHAPTER IV. Psychological Value of Geography,	23
CHAPTER V. Methods of Teaching,	36

THE BEGINNINGS OF GEOGRAPHY.

CHAPTER VI. Direction, Distance, Form, and Color,	41
CHAPTER VII. Climate, Evaporation, and Condensation,	51
CHAPTER VIII. Soil, Vegetation, and Animals,	63
CHAPTER IX. Importance of Labor,	78
CHAPTER X. Map Representation,	87
CHAPTER XI. Analysis and Synthesis,	92
CHAPTER XII. Value of Maps and Pictures in Teaching Geography,	105

THE IMAGINARY EXCURSION AND ITS PLACE IN TEACHING GEOGRAPHY.

CHAPTER XIII. A Trip Down the Hudson River,	115
CHAPTER XIV. A Trip Down the Rhine River,	132
CHAPTER XV. A Trip to Ceylon and India,	143
CHAPTER XVI. A Trip to Ceylon and India—Contd.	158

PREFACE.

The most of these chapters have appeared in *The Public-School Journal* from time to time. Some friends have thought them worthy of being put in a more convenient form, so here they are. As the title implies, they are simply intended to be suggestive in their character. No book is of very high value unless it suggests better things to the user than it contains. It is believed that this little book will do so, and that it will therefore be helpful to teachers and pupils.

The attention of teachers and of those fitting themselves to be teachers is called especially to the introductory chapters. They may not approve of all that is said; but as honest students they will seek after the truth for its own sake, and these chapters may guide them in the way that leads to it.

The chapters under Imaginary Excursions are intended as models to be followed by the pupils. It is hoped they may take as much pleasure in studying them as the writer did in their preparation. If they do, it is certain that those lessons will not prove burdensome in the least.

It is with the hope that it may help teachers and pupils to see the profit and pleasure there are in the study of Geography that this volume has been prepared by   Their friend,

HENRY MCCORMICK.

INTRODUCTORY.

CHAPTER I.

WHY CLASS GEOGRAPHY AMONG THE SCIENCES?

Some teachers and writers refuse to consider geography a science. They insist that it has no basal idea of its own on which to stand, and furthermore that it consists of facts drawn from many sciences such as astronomy, climatology, botany, zoology, history, and sociology.

It must be granted that much that properly belongs to these subjects may be, and usually is, classed under geography. And it is greatly to its credit that it can take materials from so many sources and unite them into a symmetrical whole, but if "science is the systematic arrangement of the laws of phenomena,"* the writer insists that geography is a science,

Place the Foundation. that it rests on the idea of place as a foundation,† is built up of facts which are peculiarly its own and which, with few exceptions, may be acquired in the home neighborhood, that these facts are firmly held in place by the relation of cause and effect, and that it has for its central idea, connecting it with man and his interests, what may be termed, earth-life, or the life of the globe.

Not only is the idea of place the foundation of geography, it is also the main support of all kindred subjects and makes their close correlation with

---
*Joseph Baldwin: Elementary Psychology and Education. Page 30.
†Bain: Education as a Science. Page 272.

geography possible and intelligent. It is the leading thought of astronomy: The place of the moon with reference to its planet, of the planet with reference to its sun, of the sun with reference to the many other suns which constitute the stellar system, and the places of the different stellar systems with regard to each other and to the ultimate center of gravity.

Relation of Climate to Place. It requires no lengthy dissertation to prove that climate depends on place: Place near to or remote from the equator, place near the level of the sea, place at a moderate altitude or at great heights; place on the windward side of the mountain, copiously watered; place on the leeward side, burning desert; place in the path of the warm currents of air and water, blessed with fruitfulness; and place exposed to polar currents, a frozen waste. And if it is granted that climate depends on place, it must be conceded that botany and zoology rest on the same foundation.

Relation of Commerce to Place. Among the principal factors in what is termed sociology are agriculture, commerce, and manufactures. These are selected because they are the ones with which geography deals most. Agriculture is so dependent on climate that nothing further need be said of its dependence on place. That the centers and routes of commerce are largely determined by place is also evident. Chicago at the head of a great waterway, and New York at its foot, with Buffalo at the point of transfer from the lakes to the Hudson, testify to the fact; so does the prosperity of Duluth, that city being the nearest point at which the Northern Pacific Railroad could touch the Great Lakes. Kansas City and Omaha owe their importance to their position at the gateways to the Southwest and Northwest, respectively; while the position of Pittsburg, at the confluence of the Allegheny and Monongahela rivers, of Philadelphia at the head of navigation on the Delaware, and of New Orleans at the outlet of the Mississippi basin, assured their importance, at an early day, as receiving and distributing points for large areas of territory.

Location of Cities in Other Countries. Marseilles, because of its position at the mouth of the Rhone, and in southeastern France, was important as a commercial center long before Marius and Sulla deluged the Roman world with blood, and for centuries has had a monopoly of the French trade with the Levant. The founding of Alexandria at the mouth of the Nile forbade the rebuilding of Tyre, and doomed its site to be "a place for the drying of nets." And the commanding position of Constantinople, with the Mediterranean on its right, the Black Sea on its left, the outlet of the Danube to the rear, and the route to India in front, has always made it the desire of nations.

Chicago. The character of the surface of Illinois and the proximity of its principal river-basin to that of the St. Lawrence, at Chicago, has had much to do with creating a demand for thirteen railroad bridges across the Mississippi river. Perhaps some one will assert that the construction was due to the westward tendency of emigration. But why this westward tendency? It began back in the twilight of history, when the Celts and Teutons* left their Asiatic homes and were followed by the Slavs and Huns. These enterprising pioneers moved westward, rather than in some other direction, simply because it was easier to do so, owing to the general westerly dip of the land, and also because they could travel long distances in the same climate without meeting with any rugged mountains to bar their progress.

Other illustrations of the dependence of commerce upon place, in our own country and in others, could easily be given, but it is not necessary. Enough have been given to suggest multitudes of centers and almost innumerable routes of commerce both by land and water, all dependent on their position or place.

Manufactures and Place. To show the dependence of many of the world's great manufacturing centers upon place is not at all difficult. Philadelphia and Wilmington (Del.) are engaged in the manufacture of iron and steel products because of their proximity to extensive fields of coal and iron; and their position on navigable waters determines that much of their products take the form of steel ships. The same is true of Glasgow, Liverpool, and Newcastle. Formerly all the great centers of the shipbuilding industry were found in ports convenient to suitable forests; now, because of the transition from wood to iron and from iron to steel in naval architecture, such centers are found on the Delaware, Clyde, Mersey, Tyne, and other waters easily accessible to materials used.

---

*Many learned men consider these and all other Aryans as aborigines of Europe, but the majority regard them as of Asiatic origin.

Manufacture of Iron and Glass. Wilkesbarre is in the great anthracite region of Pennsylvania, consequently the most of its manufactured products consist of machinery and locomotives necessary in the coal traffic. Pittsburg became an important center for the manufacture of iron and glass early in its history, because coal and iron were abundant in the vicinity, suitable sand not far off, and the Ohio river afforded good shipping facilities. Because of their position Birmingham (Ala.), Chattanooga, and Lynchburg, have in a generation grown from rickety little villages into opulent manufacturing cities whose names and fame are known in the iron and steel markets of the world. Minneapolis, Richmond, and Rochester have become the noted milling points of the continent because of their position near great water-falls and in productive wheat regions.

Other Manufactures. Bangor, Grand Rapids, Oshkosh, and other cities in the great pine belt, are engaged in the manufacture of lumber and wooden ware of all sorts. While New England, partly because of its rough surface, barren soil, and abundant water power, and partly because of the genius of its people, may be regarded as one great factory turning out an endless variety of articles. And in nearly every instance it will be found that the leading centers are close to water-falls or rapids to whose presence they owe their prosperity, even though some of them at present are compelled to resort to steam power. Birmingham, Bradford, Leeds, Manchester, Sheffield, and a score of other manufacturing cities owe their reputation and wealth, if not their very existence, to their closeness to the coal fields of northern England. St. Etienne, Rouen, and Lille owe their manufacturing prestige to their nearness to the coal fields of France; while Lyons revels in its silken wealth, simply because Pope Clement V, who introduced sericulture into France, resided on the banks of the Rhone.

It is not necessary to multiply illustrations. Those given will suggest many others to the thoughtful student of geography. And if he is not satisfied with a superficial view of this subject, but is determined to trace results to their causes, he will see that position or place is the most potent factor in determining the location of manufacturing centers.

CHAPTER II.

THE CONTENT OF GEOGRAPHY.

Having attempted to show that the idea of place is the foundation on which geography rests, I next call attention to the materials which enter into the construction of the science. It has already been said that these consist of geographical ideas acquired by a proper study of the home neighborhood; although it is admitted that a few of them may not be found in every vicinity, yet the number of such is small. The claim urged here is that geography has ideas enough of its own with which to rear a substantial edifice, without receiving or taking from other studies. All foreign materials are used to adorn and beautify the structure. They add to its grace and symmetry, but are not necessary to its strength or durability. The content of geography proper is given below, in part.

Classification of Elements. The ideas which belong to land and water alike are position, direction, distance, form, surface, color, and map-representation, including sand modeling and all other methods of expressing form. Those that belong especially to the land are hemisphere, continent, island, peninsula, promontory, headland, cape, isthmus, shore, beach, cliff, plain, prairie, steppe, marsh, woodland, dale, glade, plateau, mountain system, chain, range, group, peak, crag, precipice, hill, volcano, crater, slope, watershed, valley, glen, delta, gorge, chasm, gully, and canyon.

The ideas which pertain to water are mobility, ocean, sea, gulf, bay, sound, channel, strait, estuary, currents, tides (spring and neap), ebb, flow, waves, crest, trough, rivers, formation of, source, course, right bank, left bank, wearing bank, building bank, velocity, cataract, falls, rapids, bed, tributary, mouth, vapor, evaporation, condensation, clouds, mist, fog, dew, rain, springs (hot and cold), geysers, frost, hail, snow, avalanche, ice, glacier, iceberg, and icefloe.

To the atmosphere belong fluidity, expansibility, compressibility, heat, cold, winds, (constant, periodical, variable,) cyclone, hurricane, typhoon, etesian and simoom; while weight belongs to the three elements, and motion to water and air.

This classification is not perfect. Some ideas, such as evaporation, condensation, and a few others, cling with one hand to the water and with the other to the atmosphere, and it is not easy to determine in every instance which hand should loosen its grasp. It is also probable that several ideas have been omitted, as it was thought better to omit some than to insert any that did not properly belong in the list.

Notwithstanding this intentional economy, it will be seen that the constructive imagination can find here abundant material with which to build a mental picture of any country which the pupil may be studying, and *the making of correct mental pictures lies at the base of all true study of geography.* But as a building made of bricks or stones placed loosely upon each other, without cement or mortar to hold them firmly in place, would

Use of the Imagination.

be in danger of falling and killing or maiming the occupants, so a system of geographical teachings, in which the facts, or ideas, are not held in their proper position by the causal relation, is liable to topple over, burying the pupils beneath the *debris*. The fall of the structure may not kill them, but it is reasonably certain to crush out all interest in the study, and lead them to regard it with dislike, if not with loathing.

Use of Other Sciences. It is conceded that a house consisting of bare walls and a roof, no matter how well they may be constructed, is not an inviting home. These are the essentials, however, without which no amount of furnishing or adornment would avail to shelter the inhabitants from the inclemency of the weather. But in order to make the building a desirable home, the walls and floors must be covered, and the rooms furnished with grace and elegance. So botany spreads rich carpets, beautified with lilies, roses, and violets, and bordered with groves of ever-changing colors; it also frescoes the walls in traceries of the most pleasing form and delicate tinting. Astronomy lights up the edifice and causes the contributions made by botany to appear with added splendor. Zoology furnishes animals to be the associates, friends, and servants of those who are to occupy the dwelling; and manufacture takes of the materials provided by botany and zoology, and with skillful fingers fashions them into pleasing and useful forms; while commerce stands ready to furnish anything that may be lacking. And so the house is built and furnished, a fit home for him who is to be its lord and master.

Earth Life. "But," says the doubter, "botany has plant-life for its central idea, zoology, animal-life, and unless it can be shown that geography has a corresponding central idea around which may be grouped its facts, what has thus far been said goes for naught." This condition seems hard but in reality it is not so. If life is a mutual exchange of relations,* and most scientists will admit the correctness of the definition, then the earth has life; at least it exhibits the phenomena of life, and we are justified in taking earth-life as the central idea in geography. It is not claimed that this life is the same as that of the plant or animal; but *it is* claimed that it is the basis of both of these forms and that without it neither could exist.

Evidences of Life. We are so accustomed to hearing the earth spoken of as a mass of lifeless matter, that it may sound strange to hear it classed as a living organism. And yet if it has not life it has many of the appearances of life. Its molecules have their sympathies and antipathies, their affinities even, and show them in an unmistakable manner. "The magnetic needle always points to the magnetic pole, is agitated on the approach of a piece of iron, and fairly jumps under the fire of the northern lights."* A disturbance of the equilibrium of the atmosphere at any one point causes the air in the adjoining regions to move promptly towards the center of disturbance. The common gas, used in lighting our homes, may escape from the jet and fill the room with a deadly poison instead of cheerful light, unless there is heat to seal its union with the oxygen.

*Guyot, Earth and Man.

The water rushes down the mountain side, so full of joy and gladness that it cannot behave itself soberly and sedately, as becomes a "dead thing," but youth-like it goes skipping and jumping, singing and dancing, on its way to the ocean. And the **The Water.** ocean itself, how often have I heard it crooning its low, weird song to lure the fisherman out upon its bosom? And having him in its power, have I not seen it lash itself into a fury, erect its crested billows, and dash him lifeless upon the shore; or, in very spite, bury him in some one of its dark caverns, so that his sorrowing friends could not have even the sad satisfaction of planting flowers upon his grave?

Where, in plant or animal, can be found a more beneficent or complete circulation of life-giving fluid than in this so-called inanimate earth of ours? The waters leave the ocean and are borne on the wings **Earth Circulation.** of the wind to the land. They fall upon its arid surface, heal its wounds, restore its wasted energies, and cause it to thrill and throb in every fiber of its being, until it is covered with beauty and utility. Having performed their mission of invigorating the land and transforming its dormant potencies into energizing activities, the waters enter upon their return journey. Through rill, brook, and river they course, carrying with them all deleterious and effete matter found along the way, and which, if left behind, would breed disease and perhaps lead to death. From farm, village, and city they carry away the germs of diphtheria, typhoid fever, and cholera, and enter the ocean a black torrent of venous blood, to be again sent forth a pure, arterial current.

The atmosphere, too, manifests signs of life, and in its operations shows itself a true American, a democrat of the democrats. It is opposed to all class distinctions and social inequalities. It car-**The** ries the waters to all alike, unless inter-**Atmosphere.** fered with by the sun or thwarted in its purpose by the grasping of some monopolistic mountain system. It fans the brow of peasant and prince alike, favors an equal distribution of heat and cold, and drives the deadly microbes from the hut of the laborer as gladly as it does from the palace of the trust-king. This equality it tries to preserve with moderation; but if necessary it can be as noisy as a ward politician, and as destructive as "an army with banners."

These outward manifestations of life are not the only ones that exist. A thoughtful study of the form, position, and arrangement of the great land masses, that is of the anatomy of the globe, will show a plan of life and of growth which can be discovered only by an insight into the physiological functions performed by these anatomical parts. These functions, according to Professor Guyot, are, in part, "the fitting of the earth to be the abode of man, and the theater for the action of human societies; each continent being especially fitted for the education of humanity at a particular stage of its development."

CHAPTER III.

THE EDUCATIONAL VALUE OF GEOGRAPHY.

It must not be inferred from what was said in the last chapter that the writer would exclude from geography everything that does not especially belong to the so-called inanimate earth. On the contrary he is glad to include by the term all facts of climatology, physics, botany, zoology, and any other science that helps in making the earth a fit dwelling place for man,—and man himself, "to the fashioning of whose destiny, the whole animate and inanimate creation is tributary," and without whom as the most important thought, next to God, the study of nature would have no interest to geographer, biologist, chemist, or physicist.

Geography A Science. What he does object to is the view so frequently presented that geography, in and of itself, is not worthy to be ranked as a science, and must be satisfied with being regarded as an aggregation of fragments from several sciences which it has subsidized to cover its poverty and enable it to pass as a charming and useful member of the community.

It is time such views were discarded. It may have been excusable to have held them in the past, but with the teachings of Guyot, Ritter, and other masters so accessible, there is no sufficient excuse for such ignorance at present. Through the labor of these men, geography has become an individual, using the other sciences to illustrate its individuality, and having for the central principle of its being the relation of all the phenomena and forms of nature to the human race.*

There is a diversity of opinions as to the educational value of geography. Those who belittle its value, while admitting its usefulness as a knowledge study, assert that "if there be mental exercise, and good training to be got out of the study, they are secondary in importance, while in language and mathematics these are first." "Geography," they claim, "is the one department of teaching in which mere information, as distinguished from scientific method or intellectual training, is relatively of the most importance."† But, as if conscious that these sweeping statements are unwarranted, they admit that, "though much of the result we hope to gain belongs to the region of memory only, we shall be all the better for inquiring whether there is not also room here for an appeal to the judgment and to the imagination; whether, in short, geography may not be a really educational instrument, as well as a mass of facts which have to be mastered and committed to the memory." This last statement is comforting. It shows that the critics have penetrated the fog that obscured their vision and have caught a glimpse of what geography really is. The glimpse, it is true, is a feeble one, as the skirts of the fog-clouds still hinder the view from breaking upon their sight in all its splendor. It is a great improvement

<small>*Ritter, Introduction to Comparative Geography. Page 27.
†Fitch: Lectures on Teaching. Page 312.</small>

[sidenote: Objections of the Critics.]

on their previous condition, however, and improvement is always a fit cause for rejoicing.

The committee of ten, in their excellent report, treat the subject of geography somewhat exhaustively, and on the whole judiciously. **Committee of Ten.** They regard it as of equal importance with arithmetic in the primary and secondary schools, and entitled to equal time. This may be regarded as another way of saying that it has equal educative value. But while the report as a whole is a thoughtful and valuable document, it may be said by way of parenthesis, that Professor Houston's exceptions to the finding of the majority are well taken and worthy of careful consideration. There is no good reason for making physiography a distinct department of geography. The reasoning of the majority on this point is not convincing. The advantage of carrying specialization in studies to the same extent that distribution of labor is carried in the industrial world is not very apparent, even if it is claimed that our high civilization is due more to this one fact than to any other in the economic history of the world. In both cases extreme specialization may lead to greater skill and expertness within a narrow range, but it weakens the *power* of the individual, as it unfits him to see the relation of the several parts of the work to each other and to the completed whole; and geography is pre-eminently the study of relations.

Opinion of Professor Houston. This is necessarily so, for as Ritter so happily expresses it, "It is a knowledge of the relations of things that leads to a scientific interpretation;"* and notwithstanding all

---
*Ritter; Introduction to Comparative Geography. Page 15.

that may be said to the contrary, geography lends itself readily to a scientific method of treatment; indeed, without such method there can be no true study of the subject. Consequently as regards physiography, Professor Houston's views are wiser than those of the majority of the committee. It is advisable, as he clearly shows, that the facts and processes embraced under the term should be taught as a part of physical geography in the first year of the high school work. He might justly have said that they may profitably be introduced, to quite an extent, in the grammar and intermediate grades, and to a lesser degree in the primary school. They will be so introduced by every good teacher, even though the formal arrangement of the curriculum may decree otherwise.

Opinion of Dr. Harris. Commissioner Harris, in his report on correlation of studies, regards the educational value of geography as very great, both as a means of obtaining valuable information and as an instrument for mental discipline. The knowledge which it imparts, he considers of great usefulness to the citizen in his daily life; while the disciplinary value is so great that he places geography "second only to arithmetic among the branches that correlate man to nature." This secondary position may, with justice, be questioned. We will let it go at that, however, for the present, being duly thankful that geography is permitted to stand so close to arithmetic, which has heretofore been the autocrat of the common school curriculum, but which, if the signs of the times are not misleading, will soon be compelled to assume a more modest demeanor.

The importance of geography as an introduction to the study of climatology, botany, and zoology, was discussed in the previous pages, and needs no further elaboration. It was shown that it is the soil in which their roots are firmly imbedded, and from which they draw the principal share of their nourishment. It is difficult to see how there can be a scholarly treatment of the flora and fauna of a country without careful attention to its geography.

Bearing upon Manufactures. The bearing of geography on manufactures and commerce has also been pointed out, and it can be shown that its influence upon agriculture, upon which all culture depends, is equally as great. And geologists admit that its connection with geology is so intimate that it is next to impossible to tell where the one leaves off and the other begins.

No study has in itself the entire end and aim of its being. Every subject in a well-arranged school course is valuable not only because of the facts it contributes to the general stock, and of the power acquired in obtaining and properly relating those facts, but also because it leads to one above and beyond itself. It is a prophecy of something higher; and unless it prepares the pupil to realize the prophecy, it fails of its purpose. Geography is generally admitted to be an excellent preparation, not only for the subjects already named, but for the study of history as well. It is true that the facts of history can be learned without the aid of geography, but unrelated facts do not constitute knowledge, or if they do it is not abiding knowledge. There may, however, be other relations than that of place, yet the drama of history requires a suitable stage for its acting. This stage is the earth of which geography is the description. There is too much teaching of history up in the air; and although cloudland may be a fit dwelling place for the dreamer, it is not for the student.

Influence of Geography Upon History. Many illustrations might be given showing the influence of geography upon history, a few must suffice. It is a remarkable fact, pointed out by the late Professor Guyot, that the great civilizations of the world all originated in the northern continents, and that as far as known no civilization worth mentioning originated in any of the southern ones, unless it be that of Egypt; but that in all probability was an overflow from Asia. He ascribes this fact largely to the character of their coastlines.

Influence of Contour. The contours of the northern continents are irregular. Great peninsulas extend into the oceans and are bathed in vapor which renders the soil fertile and thus causes it to produce bountiful harvests. Gulfs and bays penetrate the land, forming gateways by which the life-giving fluid reaches well into the hearts of the continents. These indentations not only form gateways for the moisture, they also become common fishing grounds for the inhabitants along their shores. Casual meetings when in pursuit of their scaly prey lead to a comparison of boats, nets, and fishing appliances in general, which in turn leads to an improvement in the appliances of all the parties. Barter soon springs into existence, and in the course of time becomes worthy of being called commerce. This leads to a fuller and more general exchange of relations by which the conditions of the people are improved, both physically and intellectually, and civilization advances with sure and steady steps.

Their irregular coastline may be the cause of this advancement, or it may not; but it is certain that there is, at least, a remarkable coincidence between the contour of the continents and the character of their indigenous civilizations. Europe, the most irregular of the continents, has played the most important part in the world's history, and it is there that intellectual thought has reached its high-water mark. Greece is still the schoolmaster of the world, especially in philosophy, poetry, and sculpture. And those who would gain renown in the departments of metaphysics and pedagogy hasten to the land of Kant, Hegel, and Herbart.

Asia, too, had its civilizations and performed a notable part in history. Not as brilliant as that performed by Europe, neither is its coastline as irregular; while Africa, with the most regular outline of all, is still the "dark continent." The two Americas and Australia had their civilization imported, and the success of the transplanted article is dependent somewhat on the original stock. Yet the continent of the north has already demonstrated its superiority over those of the south.

Influence of Relief. Relief is also a prominent factor in the civilization of a nation. The early preeminence of Greece was not due entirely to the character of its coast line, or to the inherent genius of its people; it can be traced in part to the fact that the country was traversed in various directions by ranges of mountains far enough apart to leave room between them for the rise and growth of small states. The mountains were sufficiently high and rugged to form natural boundaries, but not to hinder communication.

Each state could see what the others were doing, and, being determined to excel, a spirit of emulation was begotten that carried these city-states to the pinnacle of ancient civilization. So many are the illustrations that might be given showing the importance of contour and relief in determining the historic standing of nations, that the bare recital, it is feared, would weary the reader; in mercy, therefore, most of them are omitted.

Influence of River Basins. Without a careful study of the relief and drainage of France, the student of history will find it difficult to understand why the basin of the Loire has been the theater of so many great events. Here Cæsar besieged Avaricum, and notwithstanding the stoutest efforts of Vercingetorix, captured the city and slaughtered forty thousand of the inhabitants; here Attila, at the head of five thousand Huns, was stopped, defeated in a great battle, and compelled to retreat to the swamps of Hungary; here Clovis defeated Syagrius, and firmly established the power of the Franks; here stood the Christian and Moslem, face to face, when the blows of Charles Martel sent the broken remnants of the Saracenic hosts reeling across the Pyrenees, after leaving three hundred thousand of their slain on the field of battle; here the Black Prince defeated the French and captured their king; and it was here that Joan of Arc compelled the English to raise the siege of Orleans, an event that led to the crowning of the Dauphin and to her own execution as a witch.

This famous river-basin was the scene of other stirring events, especially in the wars between the Catholics and Protestants; but enough has been given to cause the thoughtful student to inquire why this seemingly out-of-the-way corner of Europe should have been the battlefield of Roman and Barbarian, of the cross and the Crescent, and of the Anglo-Saxon and the Gaul? He will find the answer in the position of this basin with reference to those of the Garonne and the Seine, which caused a portion of it to be a part of the highway between northern Europe and Spain.

The teacher of history is often called upon to answer such questions as these. "Why was it that Washington's forces at Morristown, although not daring to face the British, yet compelled them to go by way of Chesapeake Bay in order to reach Philadelphia?" "Why was Grant at Cairo able to threaten the lines of the Confederates from the Mississippi to Nashville, and even farther?" And "how did it happen that Johnson with his comparatively small army caused Sherman so much trouble in his march from Chattanooga to Atlanta?" The answers to these and to all similar questions are found in the geography of the several localities.

CHAPTER IV.

PSYCHOLOGICAL VALUE OF GEOGRAPHY.

So far the educational value of geography has been pointed out only as it serves as an introduction to the intelligent study of other subjects. It would be a mistake, however, to infer from this that its worth is confined to the service it renders as handmaiden to others, however useful it may be in that capacity. It will be found, on a thoughtful consideration of the subject, that it has a high psychological value; that as an instrument for mental discipline it is unsurpassed.

Perception. [It affords an excellent opportunity for the training of the observing powers, as in its early stage it deals entirely with objects that can be found in nearly every community, and that may be seen by the young tyro. For this reason it should appear early in the course, when the senses are most active and eager to grasp everything that comes within reach.] Unfortunately beginners are not always set to studying geography itself, but to the study of words and maps which are merely symbols of geography. A book is placed in their hands before they are prepared to put meaning into what it contains. Or if the teaching is oral it is of the straight-jacket variety. The pupil is compelled to express his thoughts, if he has any, in phraseology which the teacher has copied from some book and written on the blackboard, instead of being permitted the freedom of expression which is so natural to children and which should characterize all teaching in the primary school. This course smothers all spontaneity, all interest, and the child becomes truly, "a chip of the old block." Geography is not to blame for this sad condition of things, neither is the child. The blame properly belongs to the teacher, who is so occupied in trying to hide his ignorance of child nature and of the nature of the subject which he is trying to teach, that he has no time to study either the one or the other.

Memory. [Geography is also useful in exercising and strengthening the memory.] Some teachers belittle the office of memory, and in so doing consider themselves worthy of praise. This is a mistake and arises from a misconception as to what memory is. They either regard memorizing as synonymous with "learning by heart," a process which "may be entirely sensuous, and which often produces weariness of mind rather than mind activity;" or else they consider the memory a receptacle, a kind of storehouse in which all sorts of odds and ends are gathered, as old furniture, unhinged trunks, and superannuated articles of clothing, are stowed away in the garret. This, if worthy of being called memory at all, is what may be termed verbal memory, and is of little worth, especially if there are no thoughts back of the words.

But that power of the mind by which one is enabled to hold related thoughts in their proper setting, so that when the occasion for using them arises, they can be re-collected and made to appear promptly, accompanied by their relatives, instead of making a tardy appearance an hour or two after, when they are not needed, must be admitted to be an important stage of mental development.

Kant's Testimony. Kant declares memory to be the chief auxiliary of the understanding,* and conjointly with observation it lies at the base of all mental development. It enriches the learner with the wealth of the past, and enables him to enter upon his inheritance, and use it in acquiring still greater treasures, instead of being a pauper confined to the narrow limits of the present. It is a pitiable sight to behold a man who has traveled far, read much, and perhaps studied profoundly, and yet is not able to recall his facts or conclusions when he needs them most.

Furthermore, what can one whose memory is feeble from the lack of proper exercise, or defective from any cause, have to reflect upon? For, as Professor Sully so clearly expresses it: "Unless the mind is stored with a good stock of concrete impressions there will be no materials for the imaginative or inventive faculty to combine, or for the understanding to reduce to general concepts."† But they are not in the mind unless they are available for use, hence a person with a very feeble memory has but scanty food for thought. He may appear thoughtful, but in reality he is in a reverie, and his thoughts are not worth a penny. The value of geography in exercising the memory lies in the fact that it deals with related truths, or ought to, and that these truths are

---

*Quoted in Sully's Hand Book. Page 172.
†Teachers' Hand Book of Psychology. Page 172.

Psychological Value of Geography. 31

associated with objects that lie in the pupils' field of vision.

Imagination By the proper study of geography the imagination is strengthened and trained. And it is thought by some that if it performed no other function than that of widening and deepening this power of the mind, it still would be well worth studying. Johonnot regards imagination as "a highly practical faculty, the one which more than any other enables man to master the forces of Nature, and raise himself above the domain of sense." He considers it "the moving force in every step of human progress, by constructing ideals which are higher and better than any that have yet been realized."*

If these statements are accepted as true, it must be admitted that any study which tends to keep the imagination from becoming morbid and unhealthy by dwelling too much upon the emotions, and tones it up, as it were, by fixing it upon beautiful and varied forms, is of great educative value. This geography does in a marked degree. The constructive imagination, the artist to whose skill and fidelity we owe most of our geographical knowledge, is kept busily at work. Its office in this connection is to build correct images of the unseen from what is known of the seen; and since but little of the world is seen by children, the accuracy of their knowledge concerning it must depend largely on this image-making power; hence the need that it be properly exercised and trained.

They may never see the Ganges river nor the

---

*Principles and Practice of Teaching. Page 12.

stately temples that rear their domes and minarets above its placid waters, yet we do not wish to have them grow up in ignorance of either. So we aid them to build a Ganges of their own from the geographical facts acquired by the home stream, and adorn its banks with religious edifices erected out of the notions obtained from the study of the home church. Imagination takes these concrete notions, and modifying them by what is read, by the instruction given by the teacher, and aided by maps and pictures, builds them into ideal forms. It is in this manner that we acquire the greater part of our geographical knowledge; hence the imagination is kept constantly active. It is active, too, in dealing with objects and their relations: a fact which compels it to take upon itself a certain degree of sobriety and moderation instead of indulging in wild flights of fancy.

Reflection. Leading pshychologists tell us that "The detection of similarity and diversity is the fundamental activity that underlies all thinking." If this is true, the helpfulness of geography in building up the understanding must be evident, as a true study of the subject involves constant comparison by means of differences and resemblances. From the same source we learn that "Inquiry into the cause of things has always constituted a chief part of human investigation." In the study of no subject does the question, "Why?" present itself more persistently than in that of geography. It will not down until it is answered intelligently; and woe to the student who tries to dodge it. Leanness of soul will be his portion, and dissatisfaction his insepar-

Questions Suggested. able companion. It is Why? Why? constantly. Why is not the state of Nevada as well watered as California? Why are the deserts of equatorial South America on the west side of the Andes, while those in the southern part are on the east side? Why are the most prosperous countries of Australia in the eastern and the southeastern parts of the continent? Why do the trade winds blow toward the west, and why do the simooms change their course with such regularity? Why is it that in the three northern continents the types resemble each other so closely, both in the vegetable and animal kingdoms, that only the practiced eye of the scientist can detect any differences, while in the southern continents the types have almost ceased to have anything in common? Why do the primitive races of the Old World present such marked differences; while those of the New resemble each other so closely that high cheek bones, copper color, and straight, black hair, are characteristic of all Indians from the Arctic to the Antarctic ocean?

Testimony of Dr. Harris. These questions, selected at random, are only a few of those that are ever in the path of the student, yet they are sufficient to show that geography furnishes abundant opportunity for tracing effects to their causes, and so may be made very helpful in training the judgment. It will be found so helpful that I feel justified in closing this division of the subject with a sentence from Dr. Harris. "What educative value there is," he says, "in geology, meteorology, zoology, ethnology, economics, history, and politics is to be found in the more profound study of geography, and to a proportionate extent, in the study of its merest elements."

Refining Influence of Geography. There is one other point on which a few words should be said. They are needed, as the point is usually ignored by the teacher, although of great importance to the children; I refer to the refining influence of geography. No subject does for the pupil what it should unless it leaves him better, as well as wiser, than he was before entering upon its study; and there is nothing that more surely wins to goodness than conscious contact with the beautiful. The teacher should constantly bear in mind that he is educating human beings, and not simply arithmeticians, geographers, grammarians, or historians; and any teaching that does not touch the heart of the pupil and make it more God-like in tenderness, sympathy, and purity is, to say the least, not good teaching. Humanity is greater than scholarship, and pains should be taken to make it as beautiful as the love of the Father. The Father has willed it so, and has filled the earth with beauty and sublimity—shadows of Himself.

How Studied in the Lower Grades. The study of geography should, in the lower grades especially, be the study of a series of pictures which the changing seasons cause to pass before the eyes of the children. The bright, fresh verdure, the budding trees, and the opening petals of spring; the billowy meadows, ripening wheat, and tasseled corn of summer; the golden fruit, dark-brown oaks, and gorgeous maples of autumn; and the sparkling frost, pendent icicles, and glistening snow of winter, all have their charms. They appeal to the finer nature of the children, and call them away from what is groveling and mean in their environments. The majestic river pursuing its peaceful way to the ocean, and bearing on its bosom the commerce of the nation, and the glowing sunset tinting the sky with golden hues, intensify the yearning after the beautiful. While the gloomy forest, the roaring cataract, and the solemn mountain impress the pupils with the sublimity by which they are surrounded, and tend to lift their thoughts to Him who created the heavens and the earth, and filled them with beauty and grandeur for the benefit of His children.

Summary. From what has been said, we see that geography should be taught—

1. For the mental discipline that may be obtained from it.

2. For the knowledge it contains.

3. For its value in connection with commerce.

4. Because of the basis it affords for the intelligent study of other subjects.

5. It should be taught for its refining influence.

CHAPTER V.

METHODS OF TEACHING.

Two Methods of Teaching. There are two general methods of teaching geography, the analytic, and the synthetic, each of which in practice, admits of various modifications. Some teachers consider it necessary to begin the study with a general view of the globe, in order that the pupils may see the relations of the different parts to each other, and of each part to the whole. Others, while admitting that it is proper to begin with a whole, declare that the whole should be a hemisphere, or a continent, as the entire earth is too large for the children to form a proper concept of it, at this early stage of their progress. The criticism is partly just, but it is fully as difficult for them to form a true concept of a hemisphere, a continent, a country, or even a state. In any of these cases we can only hope for an approximately correct concept to be formed. So that the objections against making the earth the initial whole in teaching geography are just as valid against any other whole which embraces more than lies within the children's field of vision. My own experience with children leads me to believe that the earth as a whole is more easily comprehended by them than is any large portion of it taken as an entity.

The Synthetic Method. Still others think the proper way is to begin with a small portion of the earth, that which the children can see with their bodily eyes; that they ought first to be made familiar with this in order to sharpen their vision for the later geographical conceptions, and their intellect for the more complicated relations; and that the earth as a whole should be considered only in the higher grades of study. These teachers advocate the synthetic method, which they claim is in accordance with correct principles of pedagogy, in that a small and easily comprehended space is treated at the outset; that the most concrete things, easily understood by the children, form the groundwork of further instruction; and that the gradual extension of these small districts is well accommodated to the gradual development of the pupil's mind.

The True Method. The fact is that each of these methods has its advantages and disadvantages, and the teacher who uses either to the exclusion of the other is not wise. In preparatory geography, that is, in the work which may and should be done before the children begin using the text-book, the synthetic method should have the field almost entirely to itself. But as soon as they have acquired a good stock of geographical concepts from the study of the forms in the home neighborhood, and are ready to begin the study of regions which lie beyond their field of vision, the analytic method will come into use. It is not wise to follow the synthetic method too far. There is danger that it will lead the children blindly from parts to a whole, keeping them in suspense as to the outcome. And this is to be deprecated notwithstanding the pedagogical dogma, that we should go from the known to the unknown.

The Beginnings of Geography

CHAPTER VI.

DIRECTION, DISTANCE, FORM, AND COLOR.

Need of Knowing Children's Knowledge of Geography. Before entering upon the systematic teaching of any subject, the teacher should determine, as far as possible, the extent of the children's knowledge concerning it. This he must do if he expects the best results from his labor. What they already know bearing on the matter in hand is the working capital, the use of which the teacher must so direct that it will bring in the largest possible gains for the time and effort expended. He should ever bear in mind that all he can do is to direct, to influence, to surround his pupils with the proper environment. He can, by his manner and by the interest with which he imbues the subject, stimulate them to put forth effort, and that is about all he can do. **What Do the Children Know?** All advancement, all growth, must be the result of their own effort, guided intelligently by the teacher, and this intelligent guidance can only follow a knowledge of the mental content of the learners.

Perhaps it would be better to call what the children know of the subject, when about to enter upon its study, the foundation on which the teacher must **No Sudden Transition.** aid them to erect their educational edifice. The transition from what they know to what they are about to learn should be so gradual as to be well-nigh imperceptible even to themselves. This must be so if the children are to retain in school the naturalness, the feeling of ease and familiarity, when speaking of the subject, that characterized them before entering school. A sudden transition from the familiar to the strange dazes them, and often makes them so timid that they will say nothing, for fear of not saying the right thing. This timidity leads to awkwardness and a rigidity of manner which frequently bring upon them the charge of stupidity—a charge of which, in justice, they should be acquitted at once. But whether we regard their knowledge as their working capital or the foundation on which they are to build (the thoughts are not far apart), the fact to be emphasized is that the teacher should have as full a knowledge as possible of its amount and kind.

It may be necessary to give the children some instruction on direction, distance, form, and, color; if so, the following plan is suggested. But if they show that they already know those subjects, the teacher should pass them by.

Direction. Many people fix direction by streets or railroads, and so are "turned around" when they go to places in which the streets or railroads extend in other directions than do those at the home town. The teacher should help his pupils to fix direction by the sun, and urge them to notice its position whenever they go to a new place and before deciding what is north, west, etc.

If the children do not already know the *cardinal points*, they should be taught them. The best time to introduce the subject is a few minutes before noon. If they stand at this time with their backs

Cardinal Points. to the sun, they are looking *north*, and behind them is *south*. This is true at all seasons of the year, while it is not true that the sun rises in the *east* and sets in the *west*. If they stand with their faces to the sun, they are looking south, and north is behind them; they thus learn that north and south are opposite each other.

When looking to the north, if the children stretch their arms straight out from their sides, their right hands will point to the *east*, and their left to the *west*. But if they stand facing the south, their right hands will point to the west, and their left to the east; thus they learn that east and west are opposite each other. The teacher should lead the pupils to see that, by a knowledge of direction, they are enabled to fix the location of objects and places with reference to certain fixed points, as well as with reference to their own position.

It is well to drill frequently and carefully. Drill work is valuable not only in testing the pupil's knowledge, but also in clearing it of mistiness, and in fixing it firmly in his mind. The following drill is suggested at this point; "Point to the north;" "to the south;" "to the west;" "to the east." "You may face the north." "When facing the north, what point on your right?" "On your left?" "What point opposite the north?" "Opposite the east?" "The south?" "Name several objects in the room north from you;" "several south from you;" "west;" "east."

Semi-Cardinal Points. When the teacher is satisfied that the class knows the cardinal points, let him begin on the *semi-cardinal points*, but not before. He should call attention to the fact that half way between north and east is *northeast*, so called because of its position. By a little thoughtfulness and skill on his part the children will find out for themselves that *northwest* is midway between north and west; *southeast*, midway between south and east; *southwest*, midway between south and west.

Here again drill work is in order. It should cover not only the semi-cardinal points, but the cardinal points as well. The questions by the teacher should not only test the children's knowledge; they should also be of such a character as to lead them to see the points in their proper relation. He may make a diagram on the floor, showing both the cardinal and semi-cardinal points in their proper places. After calling the attention of the children to the diagram, let him erase it, and ask each of them to reproduce it on his slate. Eight of the children may stand on the floor, arranging themselves so as to occupy the eight points. Let the teacher name any two of these points, asking the pupils occupying them to exchange places.

A diagram showing the points should be placed on the north wall and the attention of the pupils called to the fact that the north is towards the top. Other devices should be tried. Every successful teacher must be an inventor.

Distance. Next introduce the idea of distance. To learn distance it is necessary that the children notice carefully certain measures. Each should be provided with a measure a foot long, as the foot will be found the most convenient unit with which to begin. The teacher should bear in

Direction, Distance, Form, and Color.   45

mind that the important thing to be done here is, not to teach the children that a certain number of units of one order make one unit of a higher order, but to help the children to form correct mental pictures of the distances represented by the measures used. The following plan of work is suggested; but the teacher must feel free to modify or discard it altogether, as there can be no successful teaching without freedom in the choice of methods.

Method. "This ruler which I hold in my hand is a *foot* long. You may hold your hands a foot apart. I find by measuring that John has his hands too far apart, and the rest of the boys have theirs too near together. Mary has her hands about right, but the other girls have theirs too far apart. You may try again. I find now that two boys and three girls have their hands the right distance apart, but most of the others have theirs too far.

"Point to some object in the room a foot long. Name as many such objects as you can, and measure them to see if you are right.

"You may all pass to the blackboard and draw a line a foot long. Each of you measure his own line. How many have a line of the proper length? Whose is too short? Whose too long? Erase those lines and draw three more lines each a foot long. Measure the lines with your rulers; whose are of the proper length?

"The rulers that you hold in your hands are marked off into equal spaces by means of lines drawn across them. Each space is an *inch*. How many spaces are there? Then a foot equals how many inches? Your pencils are how many inches long? How long are your slates?

"Draw a line on the blackboard three times the length of your rulers. This line is a *yard* long. Look at it carefully. You may erase it. Point to some object in the room that is a yard long. Name two objects that are a yard apart. Place your ruler one yard from your own desk. This room is how many yards long? How many yards wide.

"Pass to the blackboard, and without using your rulers draw three lines, each a yard long. Measure them to see how near right they are.

"Cora, you may stand one yard from me; two yards; three yards; four yards; five yards; move half a yard further off. James may measure the distance and see if Cora guessed right; if so, she is a rod from me, as five and one half yards make a rod. How many rods in the length of the room? In the width?"

It will be well to have in the room a foot, a yard, and a rod measure. As by frequently looking at them, the children will be aided in forming proper concepts of the different lengths.

It is not necessary to teach the idea of a mile at this stage of the work, as the children will have no occasion to use it for some time yet. When it is taught, it should be by calling attention to two familiar objects that are a mile apart, and that can be seen from the school house.

The teacher will notice that throughout this illustrative work the children are urged first to guess at the length of objects, or their distances apart, and then to measure them. Why ask them to guess at the length of objects and distances which they are expected to measure?

Geometrical Forms. It is possible that such elements and figures as are necessary in map representation may have been learned in the drawing class, and the children may now be able to recognize, to name, and to represent them readily. Vertical, horizontal, slanting, perpendicular, and curved lines; the right angle, acute and obtuse angles; the varieties of triangles and quadrilaterals, and the circle, should be familiar to them before they begin to make maps. The practice in making these figures will give the learners confidence in their own ability, and strengthen their power to detect and learn the different forms.

In teaching this subject, it will be well for the teacher to stand with crayon in hand as he talks to his class and place the elements and figures named on the blackboard. The following method is suggested:

"This straight up and down line is a vertical line. Draw three vertical lines on your slates. Draw three vertical lines, each six inches long. You may pass to the board and draw five vertical lines, each a foot long. Point to several vertical lines in the room.

"This is a horizontal line. How does it differ from a vertical line? Draw on your slates two horizontal lines the length of this one. How long is this line? Draw on your slates four horizontal lines, each four inches long. Draw six horizontal lines on the board, each a foot long."

In like manner right-slanting, left-slanting, parallel, straight, and curved lines may be taught. When teaching slanting lines, the teacher should introduce the word "oblique," and show that it means the same as "slanting."

An *angle* is the next simplest form after the line. It is formed by the coming together of two straight lines, and is their difference of direction. Let the teacher show his pupils what this means, pointing out to them that the size of an angle does not depend on the length of its sides, but on the difference of direction. The right angle being the standard of measurement, should be introduced first, then the acute and obtuse; the meaning of those terms should be explained. This is also a good time to make clear to the children the difference between vertical and perpendicular lines. It will be difficult to convince them that a horizontal line may be a perpendicular line, and that a vertical line is not necessarily a perpendicular line, yet it must be done.

The transition from the angle to the *triangle* is an easy one, there being but the addition of one line. It should be pointed out, however, that this slight addition gives two additional angles; and that the completed figure owes its name to the fact of its having three angles.

An addition of two lines to the triangle gives a figure having four sides, and every four-sided figure is a *quadrilateral*, as this long word means that which has four sides. If the angles are all right angles and the opposite sides are equal, the figure is a *rectangle;* and if the angles are right angles and all sides equal, it is a square.

Let the names of the figures be taught in every instance, instead of having resort to some round-about expression, which is fully as difficult and far less satisfactory.

A *circle* should next be described on the board, and attention called to the bounding line or *circumference*, the *radius*, and *diameter*. It must be impressed upon the pupils that the *space* enclosed by the circumference is the circle. This is important. Definitions learned now should be good for all time and in all studies in which the terms defined occur; and there is no good reason why they should not be.

Desk Work. The teacher should not only drill the children in making those forms on their slates and on the blackboard during the recitation period; he should also be prepared to keep them occupied while at their seats. A box of colored sticks that may be bought for a few cents, will be found very serviceable in this work. These the children can arrange in the forms of lines studied, and with them they can make the different kinds of angles, triangles, and quadrilaterals. If the sticks cannot be procured, the children may be kept busy folding papers into the various forms.

Color. As stated elsewhere, the making of correct mental pictures lies at the basis of all true study of geography. Color is an essential of pictures, hence it is taught as an aid to seeing and conceiving forms as they appear in nature. It is possible that this subject has already been taught in a special class in form and color. If so, it will do no harm to review it here; and if it has not been taught, it is time it were.

It is best to use but a few colors at first, and those as near the standard as possible. Colored tissue paper is very convenient for use in teaching this topic. "Holding blue paper towards the light, and placing with it red, violet is produced. Likewise, red and yellow give orange; blue and yellow, green. Placing the green and violet together results in olive; orange and violet produce russet."

Let the teacher place upon the table a collection of bright colored objects, such as ribbons, fruits, flowers, balls of yarn, etc., and have the children name the color of each object, and place objects of like color in a group by themselves. They should be required to select some object on the table, and name all objects of like color in the room; also to observe the different colors in clothing, leaves, flowers, fruits, animals, the rainbow, etc.

A box of ordinary water colors will be of service to the teacher in doing this work; but if each pupil is provided with a box with which to paint familiar objects in brilliant colors, success is certain.

CHAPTER VII.

CLIMATE, EVAPORATION, AND CONDENSATION.

Forms of Land and Water.
Children, by the time they are six years of age, have acquired a large number of geographical facts, and can reason about those facts more intelligently than many suppose. Their earliest knowledge of geography is along the line of physical geography, close to which follows sociology, while political geography brings up the rear. Until quite recently the text-book makers tried to reverse this order, and some teachers unwisely followed their lead. The results were usually unsatisfactory; they could not well be otherwise. There is no more crying evil in connection with the teaching of geography than this, that the children are torn away from Nature with whom they have grown into close companionship, and compelled to study the works of man, and, at first, usually those of least interest. Not but what the works of man are well worth studying, but everything in its place. First let them study Nature near whose heart they dwell, and to whose every mood their souls respond, and the habit of study formed here in dealing with that in which they take delight will enable them later to turn to the works of man with pleasure instead of indifference or even loathing.

They know what is meant by heat and cold, and that summer is hotter than winter, although they may not know why it is so. Rain, hail, and snow are familiar to them, and they understand that all three are connected in some manner with the clouds, but how connected they do not understand beyond that they drop from them to the earth. If they live in the country or in country towns, they have seen dew and hoar frost, and know that cold has caused the frost, but probably will not know it has caused the dew also. They have romped so much with the wind, chasing it when it carried off their hats, pouting when it blew their hair about their faces, and smiling when it painted roses on their cheeks, that they will need no formal introduction to it at this time; but they possibly will to the air, which has behaved itself so quietly that it has not attracted their attention.

Difference in Temperature. Knowing well the amount and character of the children's knowledge on these points, it ought not to be difficult for the teacher to lead them to see that the greater heat of summer is due to the greater altitude of the sun and to the fact that it then spends the larger part of each day, of twenty-four hours, where its heat reaches us directly. Being more nearly overhead in summer than it is in winter, the rays of heat from it come more nearly straight down, or vertical, and consequently more of them will fall on any given space, This the teacher can easily illustrate. The relation of the wind which they know to the air which they do not know, and how a difference in temperature between two places causes the wind, can be taught successfully to these young learners, if the teacher has mastered the topics himself, and has

Climate, Evaporation, and Condensation. 53

learned the art of leading children into the light. A few simple illustrations with which the children are familiar, such as the "steam" from the boiling teakettle, the steam in the laundry on wash days, and the rapid drying up of water spilt on a hot stove, will enable them to grasp the thought of evaporation; while the drops which form on the under side of the lid placed over a kettle of boiling water, on the windows in the laundry, and sometimes even on the walls, will lead to an appreciation of the subject of condensation. The pupils will see that evaporation is caused by heat, condensation by cold; that these processes which, on a small scale, are going on in and about their homes are taking place on a larger scale outdoors; and that the sun is the source of heat.

Evaporation.

It may be necessary at this point to call the attention of the children again to the air, or atmosphere. This is not an easy subject to teach to beginners, yet, with skill on the part of the teacher, they can be led to comprehend it fairly well. They can be taught its nature in a general way, its height, its weight, how it is affected by heat and cold, and how necessary it is to all forms of life. It is heavier than vapor of water, and when the heat changes some of the water of the oceans, seas, lakes, rivers, etc., into vapor, the vapor being lighter than the air is forced up by it, and at the same time carried along by the horizontal movement of the air, that is by the wind, until it reaches a place where the cold is sufficiently

The Atmosphere.

great to condense the vapor into drops of water so small and light that they float around in the air.

Condensation. The clouds consist of a very large number of these little drops. If the clouds are driven by the wind into a still colder region, the tiny drops composing them are crowded into larger drops which fall in the form of rain. Sometimes the raindrops are frozen before they reach the earth, and form hailstones; and, if the cold is great enough, the little particles of water forming the clouds will be frozen and fall as snow.

Effect of Rain upon the Earth. The effect of the rain upon the earth can be taught. The heat from the sun and the moisture cause the crops to grow, the flowers to bloom, and the trees to send their roots down deep into the ground, rear their heads high into the air, and send out their strong arms to gather in and protect the birds of the air and the beasts of the field. Some of the water sinks into the ground and comes out again in the form of springs; some of it is taken up by the plants, some is evaporated, and most of the remainder is gathered into small streams, those into larger ones, which in turn unite to form large rivers that flow into the ocean, to be again changed to vapor and carried to the land.

Illustrations. The pitcher of cold water that stands on the dining-room table at the time of the mid-day meal, in summer, will furnish an illustration of the manner in which the dew is formed. Whence come the drops of water that form on the outside of the pitcher? Why do they form but a little higher than the water reaches on the inside?

Why is it that some days no drops are formed, no matter how cold the water may be? And why do they never form on a pitcher of hot water? In answering these questions, the children will use and clarify what they have already learned concerning evaporation and condensation, and the effect of heat and cold upon the capacity of the air for holding vapor of water.

The cold water in the pitcher makes its outer surface cooler than the air in the room, and the air being still, the same particles remain in contact with the surface long enough for the cold to condense the vapor which they hold into these drops. The drops are not seen much above the surface of the water in the pitcher because the upper part is not sufficiently cold to condense the vapor. Again, some days there is comparatively little vapor in the air, and if any of it is condensed by contact with the pitcher, the drops are so small as to pass unnoticed; or the air may be moving so briskly that the same particles do not remain in contact with the cold surface long enough to lose the moisture they may contain. Why drops do not form on the outside of a pitcher of hot water will at this juncture be evident to the pupils.

The phenomena just noticed are familiar to children both in city and country, and their explanation will give an insight into the manner in which dew is formed; why there is more of it formed on a still night than on a windy one, and more on a clear night than on one when the sky is overcast with clouds. If they understand about the formation of dew, they will readily see how hoar frost is formed. It will be necessary to touch somewhat upon the subject of radiation of heat by the earth, but the teacher need not be afraid to do so; for if he does it intelligently, the children will respond heartily to his efforts.

Land Forms. Attention should next be called to the land in the home neighborhood, and first to the surface forms. The plain, the knoll, the hill, if any, the valley, and all other forms that are accessible, should be inspected by the children. Descriptions by the teacher, however accurate and vivid, are not so valuable as actual observation. It is only by coming into the presence of these forms and seeing them for themselves that the pupils acquire apperceiving power sufficient to apprehend the descriptions and perceive their meaning. It is in the same manner that they obtain apperceiving material with which to master the geography of all regions lying out of their field of vision.

Slopes. Especial care should be given to the slopes, as the entire surface of the land consists of such with their lines of union and lines of separation, the lines of union being found in the bottom of the valleys where the waters from the slopes facing each other unite, and the lines of separation in the divides, or water sheds, that separate the slopes. The importance of a careful study of these forms cannot well be overestimated, as a large share of the benefit acquired from geographical study comes from being able to form and retain a clear, accurate mental picture of the physical features of the earth; which can be done most easily, perhaps, by uniting these small slopes into larger ones, and these again into still larger, until there is the ability to arrange the entire surface of the land into two grand slopes, one long and gradual, facing the Atlantic and Arctic oceans, the other short and abrupt, facing the Pacific and Indian Oceans.

The value of this mental picture will be evident, if it is remembered that man, next to God, is the central being of the universe; that whatever affects him most, is worthy of the most careful study; and that while he may largely determine his conditions, yet as a matter of fact they are determined for him, in some measure, by the geographical conditions of his locality. The most immediate benefit to the children, however, comes from seeing the relation of the slopes to the stream at their line of union. It is poor teaching that will not lead them to see that the size of the stream depends largely on the length and width of the slopes, that its course depends on the direction of the tilting of these slopes, and its velocity on the degree of tilting towards its mouth. If they see these relations between the home stream and the slopes on either side, they will be the better prepared to form approximately correct notions of unseen streams and slopes concerning which they will hereafter study.

Home Stream. By a proper study of the home stream, the pupils learn to put meaning into such terms as the source of a river, the mouth, right bank, left bank, wearing bank, building bank, course, current, velocity, bed, sand-bar, and, possibly, rapids and falls. This stream, small though it may be, will afford an opportunity for obtaining correct notions of an island, peninsula, isthmus, cape, promontory, strait, bay, or gulf. The children's first knowledge of these forms should be gained from personal observation; that is, it should be immediate knowledge as far as possible; and no person or agency should come between the learners and the real objects which they are studying.

Unfortunately the objects themselves are often ignored, and the children are introduced to the study of geography through maps, pictures, and words, and that, too, in many instances when the objects which these symbols are intended to represent can be seen from the school house door. The symbols are well enough in their proper place, but meaningless words, especially if unnecessary, are always out of place; and it is questionable if the study of maps should ever find a place in the primary school; the making of maps certainly should, but they should be made from the objects instead of being copied.

Pictures are very helpful, when the objects pictured cannot be seen; but where they can be, they surpass even the best of pictures, and should be studied first. Thirty minutes spent by the shore of Lake Michigan would give the children more correct notions concerning lakes, seas, and oceans, their waves, the surf, cliff, beach, etc., than would a week's study of the best painting of any one of them. The educative value of the picture depends on their ability to interpret it correctly; that of the object is due to its adaptability to their need.

Value of Pictures.

We sometimes fail to recognize how admirably the child and the universe are fitted to each other. The child is so small and the universe so large that it is difficult for us to believe and feel that the highest and noblest purpose of everything in the universe is the education of the child. God in his goodness has given us abundance of suitable material for the accomplishment of this purpose. But in our pride or ignorance we turn away from it all, and substitute therefor the work of our own hands, forgetting that He who created and fashioned the universe also created the child, and that possibly He knew as much as we do about what was suited for its intellectual and spiritual growth.

Work of the Water. Not only can the forms of land and water be studied near home, but here also the children can be led to understand, in a measure, the results of the action and reaction of the two elements upon each other. They have noticed that when it rains, the water flows in rivulets in the streets, or in the country roads, and is muddy. In the streets of the city the rain washes the dirt into the gutters near the curbing, where it is carried along by the water in two muddy streams. As the shower becomes lighter, these streams become smaller and clearer, and shortly after the rain ceases they disappear. Before doing so, however, they carried much of the dirt for quite a distance, and left it in patches along the sides and in the bottoms of the gutters. This power of the water to dig up the dirt and carry it from one place to another can be seen on a country road much better than in the streets of the city. After a heavy rain it will be found that the ridges are cut down, the ruts filled, and the road smooth. The children will know that these changes are due to the action of water—they knew it before, but did not consider the knowledge of any value, as the teacher said nothing about it.

If the pupils live in the country, they will see how much more rapidly the rain cuts furrows and gullies in a newly plowed field than in a pasture, or even in a field that has been plowed some time; and they know why this is so. Here, as well as elsewhere in beginning geography, or any other of the natural sciences, the teacher should constantly appeal to the experience of the children; a large part of his work must consist in organizing their knowledge and making it available in the acquisition of further knowledge.

The pupils have now learned two things in regard to the action of the water upon the land; first, it loosens the soil, and second, it carries it from one place to another. What they have seen the water doing in the street, on the road, or in the plowed field, they can see it doing in the pasture. Here the roots of the grass partially protect the soil from the loosening and carrying power of the water, so that furrows and ditches will not be made as rapidly as in the plowed field. But in the course of time many such have been made even in pastures and meadows. The creek, which they cross perhaps several times a day in coming to and going from school, was made in this manner.

On a rainy day a number of little streams may be seen flowing into it on either side. The water in these is muddy, showing that they are carrying dirt or soil into the creek; and the water in the creek is still more muddy. Both creek and rivulets have washed out the land along their banks, and the creek, being full of water, flows more swiftly than usual and carries with it the dirt. As soon as the rain ceases the water passes out of the smaller streams leaving their channels empty; much of the water in the creek also passes off. What is left does not more than half fill the creek, and it flows more slowly than it did when the creek was full. As a consequence of the small amount of water and the slow movement, its carrying or transporting power is diminished, the soil settles to the bottom or along the sides and the water becomes clear.

The processes of loosening and transporting the materials of the land from one locality to another, which the children see taking place near their homes, is taking place on a larger scale all over the land. The water is constantly wearing away the ridges and filling up the hollows. If the land is soft, the wearing is rapid; if it is hard the wearing is slow. But even the hard rocks are changed little by little into gravel, sand, and mud. In making these changes the water is aided materially, in some parts of the world, by the frost. When it rains the crevices in the rocks are filled with water. If the weather becomes very cold before the water is evaporated it freezes, and in doing so chips off a piece of the rock, as any quantity of water requires more space in the form of ice than it does as water; this is why the pitcher in the cold room is broken if the water in it is permitted to freeze. The pieces of rock chipped off by the frost are ground against each other as well as against the rocks from which they are chipped, and then carried down the stream.

Wind and Frost. The wind, too, in addition to "carrying the waters from the sea to the land as fast as the rivers carry them from the land to the sea," helps in wearing away the rocks and soil. If there is a hole in the rock and it contains a few grains of sand, the wind will whirl the sand round and round. The little grains have sharp edges, and are sure to make the hole larger and deeper, and when the rain fills it with water and Jack Frost freezes the water, larger pieces of the rock are broken off. The wind wears away the loose soil much faster than it does the hard rock. Finding a small hole in the soil, with a little dust in it, it uses the dust in making the hole larger, and then frequently blows it away to some other place, as perhaps it did the sand which it ground from the rocks.

These three destructive, yet beneficent agents—the frost, the wind, and the rain—are constantly at work, under the supervision of the sun, changing the forms of the land. Every rill, creek, and river is busy carrying materials from the highlands, where the cold may be so great that vegetation will not thrive, to the lowlands where it will. They have chiseled and carved great masses of highlands into hills, mountains, valleys, gorges, and canyons; formed the loosened dirt into fertile plains, or with it converted the commodious harbor into a fever-breeding marsh. Therefore it is entirely proper to say that the rivers formed the hills, valleys, and plains. The manner in which they did this, and are still doing it, can be illustrated fairly well by means of the molding board, a little sand, and a sprinkling pot. It is far better, however, to call attention to these processes as they are taking place in the home vicinity. Observant teachers will find in their natural environment nearly all the illustrations they need, and will rarely resort to art; when they do it is usually to supplement—not to supplant nature.

CHAPTER VIII.

SOIL, VEGETATION, AND ANIMALS.

How Soil Is Formed.
When studying the home stream we noticed that during, and immediately after a heavy rain, the water was muddy. We also noticed that shortly after the rain ceased the water in the stream became lower and clearer, and that strips of sand and mud, with perhaps quite a mixture of leaves and other woody matter, were deposited on one or both sides of the stream. What we there saw on a small scale may be seen on a large scale by large rivers, with this addition, that the deposits will contain many pebbles, especially if the rivers rise in a hilly region. Soil formed in this manner is very fertile, and is called alluvium, because formed by the action of flowing waters.

The Wood Lot.
If we examine the wood lot we shall find on the surface a quantity of leaves in a state of partial decay; under these a layer of vegetable mold, showing quite a liberal sprinkling of woody fiber; while still lower we find the soil growing harder and changing from black to a brown or reddish color, with only slight traces of vegetable matter, the bulk of it being fine sand and rotten rock. Under this will be found the bedrock, with its surface rough and uneven and showing strong signs of decay. This soil is different from the alluvial in many respects, more especially in

these two, that its richness depends partly on the nature of the rock below it, and that it was formed where now found? But how formed?

Some of the children may think that it was always as it now is; this is not correct. They should know by this time that the face of the earth is constantly undergoing change—change in both its vertical and horizontal configurations. Alternations of heat and cold, by expanding and contracting the surface of the rocks, are tearing apart the crystals. The air, frost, and rain are unceasingly rotting and loosening the rocks and transporting them from their mountain fastnesses, as if with the design of dumping them into the sea; while the sea, angry at the encroachment upon its domain, hurls itself in fury against the margins of the land, undermines the tall cliffs, grinds the rocks into sand, which it mixes with the off-scouring of the earth, and buries the unseemly mass fathoms deep, to await a future emergence from beneath the waters.

No! nothing is as it was; all things are becoming, and it is only by a long process that the earth has become what it is. It is believed that at some time in the past all the land that was above the water was hard rock. A part of this had to be broken up, pulverized, and changed into soil in order that plants could grow and support animal life, so there might be both plants and animals to contribute to the well-being and happiness of man. The manner in which the alluvial soil was formed has been described; it remains to tell how this native soil, so called because formed where it is found, was made.

Constant Change.

Introduction of Vegetation As soon as the surface of the rock was slightly roughened by the agencies already named, tiny plants appeared in every hollow in which they found a pinch of rock dust. These small plants were lichens and mosses, such as can be found growing on rocks at the present time. They prospered, spread over large areas, and assisted the rain in rotting the rocks.

The manner in which they assisted was this: The lower parts of the plants decayed last year, and the upper parts made a more luxuriant growth than they did the year before. The decaying plants contained certain acids which had a greater destroying power than did pure water. The rain, in passing down to the rock, absorbed some of these acids, and so was able to rot the rock much faster than it could without them. The decaying vegetation not only increased the destroying power of the water, but by being washed into the larger hollows it formed a soil in which grew plants having well-marked roots. These roots reached down into the loosened portions of the rock, and deepened the soil by mingling their lifeless bodies with the rocky fragments. Year by year the decaying leaves and stems were adding to the soil on top, and the rain and roots were adding to the lower surface by loosening and rotting the rocks; it was thus the soil in the wood lot was formed.

There are other soils, however, besides alluvial and native soils. We will speak only of one, the "drift." The soil of a large portion of our country does not rest upon the rock from which it was made, but on a great mass of stones, clay, sand, and gravel. These materials were

Drift.

shifted from where they were formed and deposited where they are now found. As they are too heavy to have been transported by water, and for certain other reasons, it is believed that they must have been brought from their homes by the action of the ice. The mass is so deep in some places that the roots of plants never reach the native rock, while in other places so much of it has been worn away that the rock can be easily reached.

The children should be led to see that whether the soil is alluvial, native, or drift, it has been formed by the action of heat, air, frost, and water. They should be made to feel that while it is satisfying to know how things came to be as they are, the study of the soil has a still higher value, as it is the foundation of life. The boys and girls cannot be too deeply impressed with the thought of their indebtedness to the soil.

The World God's School. This world is God's training school, and like all his works it is well adapted to the accomplishment of his purpose. It is abundantly provided with all the materials and forces necessary for the development of man's threefold nature. Land and sea unite in furnishing him with a variety of pleasing and nourishing foods. The study of the forces which are operating around him, and of the laws which govern them, affords ample opportunity for the development of his intellectual powers; while the beauty and sublimity which surround him on all sides are sufficient to awaken and cultivate in him a love for the beautiful, and cause his spiritual nature to reach out after the Creator and Maker of all, who is his Father, Teacher,

Soil, Vegetation, and Animals.

and Friend. Any other view of the purpose of the universe is unworthy and irrational. It mars the harmony which should exist between man and nature, while this view transforms all creation, man included, into a rhythmic poem.

These things being so, it follows that the thoughts of the children should at an early age be turned, not only towards the forms of land and water, with the forces operating upon them, but also towards the wealth of plant life by which they are surrounded. This is necessary to a proper understanding of what their geographies will tell later of the vegetation of remote regions. And while this is not the only or highest purpose, it is sufficient to entitle plant study to some consideration. It is possible, even, that at this early stage in the progress of the pupils it should be made the main object of the study; the scientific phase of the subject being held in the background till later. What was said about the formation and character of the soils will form a natural introduction to the study of plants, their forms, habits, and uses.

Study of Plants.

It may be advisable to begin with the food plants, and more particularly with those which the children can see growing in the fields and gardens. The more common they are, the more worthy of study they may be. The very fact that they are common goes a long way toward proving that they are good. The so-called Irish potato has been of more value to the world than all the gold and silver that South America has ever produced. Its crop of indian corn is worth far more to the United States every year than is its out-

Cultivated Plants.

put of the precious metals, while the plebeian rice plant, from the marshes of India, forms the staple article of food of the millions of southern Asia.

Lack of time will not permit a detailed study of each of the food plants, and such a study is not necessary to a knowledge of geography. A few, however, should be so studied, and perhaps the Indian corn is as good as any to begin with. Children who notice the seed before it is planted,

Indian Corn. examine it a few days later when it has "sprouted," observe the blades as they come above the ground, the stalk when it has tassled, the ear with its silk, and learn the relation of the ear, stalk, and tassel to each other, will acquire a knowledge that is both educative and "useful." When they have followed the ripened grain from the field to the sheller, from the sheller to the mill, and from the mill to the various pleasing and nourishing articles of food into which it is made, their education will have received an added relish. If they are foolish enough to disdain such wholesome food, they may feed the corn to the stock and learn that it is converted into meats for foreign shipment, or into ham and eggs, beefsteak and roasts, and milk and butter, for home consumption. Or if they do not care for any of these, they can see both corn and meats shipped to other countries, and the money for which they are sold invested in coffee, tea, sugar, tropical fruits, spices, silks and satins, laces and embroideries, etc., with which to tickle their palates or please their aesthetic taste.

Some of the other grains, with the more useful vegetables and fruits, should also be studied, care being taken to begin with those which are familiar; as in studying such the experience of the pupils can be most successfully appealed to. The impetus acquired through interest in these will lead to a habit of thoughtful observation, which is the foundation of all knowledge.

The study of the uncultivated plants in the lower grades should also be carried on along the three lines of form, habits, and uses. Botany can no more be learned out of a book than can geography. It is the study of plant life, and there is no life in the book. That contains simply a summary of what men have learned about plants, and while it is not the purpose to make botanists of the children at this early age, there is no wisdom in keeping them from studying the plants themselves, especially as they are far more accessible and interesting than the books. To keep on studying about a thing instead of studying the thing itself, ought to be as unsatisfactory as it would be to keep talking about a good dinner instead of eating it when it could easily be obtained.

Uncultivated Plants.

By studying the plants in the home neighborhood the children gather such a stock of information as will enable them to appreciate in a measure the vegetation of foreign countries, and to enter intelligently upon the study of the science of botany. The accumulation of related facts through personal observation is the work to be accomplished.

Study Home Plants.

How does the plant look? Can the children name it at sight? How do they distinguish it from others? Where does it grow most abundantly and with the greatest vigor, on wet land or on dry land? On the sunny slope, or in the shaded glen? For what is it used? Anything in its character that fits it for this purpose? How is it prepared for use? The answers to these questions, and to others that will suggest themselves to the teacher, ought to be learned by actual observation, and not from reading or hearsay, or even from the study of pictures. There is material enough in every neighborhood for the successful accomplishment of the work, as it is not necessary that the same varieties should be studied by all.

Grasses. Do the children know the most common grasses? They can be found almost everywhere, and are both beautiful and useful. They not only form the principal food of all domestic animals and of most wild ones, they also fertilize the land, save it from being stripped of its soil, and protect nations from the encroachments of the sea. Flowers are plentiful. God has strewn them in profusion over the earth. Have the children been led to notice their variety of form and color, and to admire their beauty? Do they know all that grow within a mile of the school house? Does the teacher know them? Has he consciously used them in trying to develop the love of the beautiful which dwells in the souls of his children? If not, he should at the earliest opportunity set to work to do it.

Trees. What of the trees "that climb the mountain sides to greet the rising sun, and with their green banners to wave him adieu at the close of day."—have they been studied? Have the shapes of the trunks, the arrangement of the boughs upon the trees and of the leaves upon the boughs, with the character of the bark, been noticed? Can the children tell by the shape of the leaf and its plan of venation the tree to which it belongs? Does an examination of the seed or fruit enable them to name at once the tree on which it grew? Do they know how the seeds are carried from one place to another, and how fitted to be thus carried? How many varities of evergreens do they know? Is the larch an evergreen? The pine? The fir, etc.?

Use of Trees. To show the usefulness of trees it may be well to take the one that enters into most common use. Perhaps the pine is as good as any, although the oak and maple have certain advantages. It is hoped that the teacher will not begin the work by studying foreign trees, simply because they may be remarkable for size or products. To begin with the banian tree of Hindoostan, the talipot palm of Ceylon, or the "big trees" of California would be unwise, as the children would have to depend entirely on mediated knowledge, a condition to be avoided whenever possible.

It may be that no pine forests are accessible to the children, yet it is likely that they see pine trees every day, since they are found in nearly every school district. If the teacher is acquainted with the felling of trees and the hauling of logs, let him take the children on an imaginary excursion to a logging camp. It will be helpful for them to see, even through his eyes, how the men live in those camps, fell the tall trees, saw them into proper lengths, and haul them to the railroad or the river, to be taken to the sawmill. He can interest and instruct

Lumbering. them by telling them how the logs float down the small river in the spring—sometimes forming jams that are broken up by the "drivers" at great risk—how they are stopped by the boom across the river and taken to the sawmill which is near by, or formed into rafts and floated to others at distant points. If the children can visit a sawmill and see the logs being sawed into boards, timbers, laths, shingles, etc., the sight will be of value to them. A sawmill, however, is a dangerous place for children, but they can with safety visit a lumberyard and see the various forms into which the logs have been sawed. It will also be helpful to take them where a house is building and let them see in what part of the structure the different timbers enter, being careful to give the names, such as sills, studding, joists, rafters, flooring, etc.

Other uses made of pine should be mentioned, the children being encouraged to name as many as possible. They may have heard or read that turpentine, tar, pitch, and resin are obtained from the pine tree, and the teacher will need to be prepared to tell how they are obtained. The advisability of discussing the varieties of pine is doubtful. Classification as such has no place here, although it is recommended that the work be done in a systematic and orderly way; but the order must not be made too prominent, lest the children consider it the principal thing, and thus the real purpose of the work be defeated.

Study of Animals. Vegetable life is not the only kind of life found in the vicinity of the school house. Animal life exists there in great abundance and in varied forms. And since this resembles human life more closely than does the life of the vegetable, its study will be more interesting to the children. Many children have pet animals, others would like to have them. If the pet is a calf, a lamb, or a chicken, its study may be made the avenue to the study of the domestic animals. If it is a rabbit, a squirrel, or a canary, its study ought to lead to the study of the wild animals. Here, as well as elsewhere, the study of the familiar should come first. Familiarity need not breed contempt, and will not unless the teacher keeps his pupils studying the familiar so long that they become disgusted. Some teachers do this, and the result is usually disastrous. It would not be at all strange if the boy knew more about a dog, and the girl more about a cat, than does the teacher. These animals will do very well to begin with, and the boy should be encouraged to tell all he knows about the dog's affections for his master, his faithfulness, and his usefulness; this knowledge may be profitably supplemented with stories from books, bearing on the same points. The girl may tell about the affection and faithfulness of the cat, also its usefulness. Both of the children, by a little guidance, can be led to see how admirably these animals are fitted to fulfill the purpose of their existence.

When entering upon the study of the animals found in the home neighborhood, it may be wise to have the children make a list of all the animals they know—those they have seen. Such a list will show that many of them have been close observers, and, as a consequence, know more of the habits of the animals than do some of their adult friends; it will also prove a valuable guide to the teacher in determining the animals to be studied. The animals in the following list were known to a boy eight years of age. He had seen all of them, and could name them at sight. He had seen others in the circus procession, but they are not given in the list:

Dog	Goose	Clam,
Cat,	Duck,	Frog.
Horse,	Guinea hen,	Toad,
Mule,	Robin,	Butterfly,
Cow,	Sparrow,	Caterpillar,
Sheep,	Peacock,	Grasshopper,
Hog,	Blackbird,	Mosquito,
Rabbit,	Crow,	Spider,
Gopher,	Pigeon,	Bee,
Chipmunk,	Bluejay,	Ant,
Squirrel,	Woodpecker,	Wasp,
Bear,	Humming bird,	House fly,
Raccoon,	Canary,	Bugs (beetles
Skunk,	Meadow lark,	in general).
Rat,	Fish (several	Potato bugs,
Mouse,	kinds,	Cabbage worm,
Monkey,	Snake (several	"Fishing" worm,
Chicken,	kinds),	Firefly,
Turkey,	Crayfish,	Miller,

Method of Study. These animals, with a few exceptions may be found in almost any country neighborhood, and many of them can be found in the crowded city—in the parks, on the shade trees, and on the grass plats that border many of the streets. And while it is true that in studying nature, the country pupil has some advantage over the city pupil, it is also true that in some respects the latter has the advantage. The public parks and gardens afford him an opportunity to study many plants and trees that are not found in the country district; while ready access to the menagerie and zoological garden enables him to study animals which the country pupil can only see in connection with the circus, and the circus is not always accessible. The fact is both city and country have sufficient material, though differing in kind. A slight difference in the forms of life is of little consequence. The important thing is that the children be led to study such animals as they can find, and note their relation to the environment.

The study of the science of zoology is not the purpose here, yet this work should prove an intelligent introduction to the study of the science at the proper time. It will not answer this end if the time is spent in studying *about* animals instead of studying the animals themselves. Even at the risk of wearying the reader by the oft repeated admonition, it must again be urged upon the teacher that unless the children see the animals they are studying, the value of the work will be reduced to the minimum. The above list, while not containing all the animals that may be found in almost any neighborhood, is full enough to show that there is no dearth of material.

As this book is intended to be suggestive simply, no course of study is indicated, as, to be of the greatest worth, the work must be laid out by each teacher for his own school. In doing this, it is probable that he will plan to have the cow and horse studied next after the cat and dog.* They should be studied with reference to form, color, size, movements, and use-

---

*McMurry's Special Method in Science is an excellent guide in this study; published by Public-School Publishing Co., Bloomington, Ill.

fulness to man. The chicken may be studied first as a type of fowls in general, and then compared with the duck or goose; the leading differences will be noticed by the children. In studying the wild birds, those injurious to fruits should be named, although it is doubtful if there is any bird that does not pay for what fruit it eats or destroys, by its destruction of hurtful insects.

Perhaps the study of no other form of life will afford so many pleasant surprises as will that of insects. Almost every neighborhood, city, **Study of Insects.** or country, teems with such forms, and many of them are very beautiful. All insects do not injure fruits or vegetables, and one of the purposes of this work should be to point out those which do. That was one of the objects of the legislators who made the law requiring the teaching of elementary zoology in the public schools. But teachers and county superintendents, by setting the children to learning long lists of animals which they have not seen, and many of which they never will see, have made worthless the good intentions of the lawmakers.

If this subject is treated as it should be, it will lead the children to be kind to animals. There is need of this. Some boys are young savages, not having yet reached even the barbarous state. They love to inflict pain simply to see the suffering they can cause. This is wrong. No one has a right to inflict pain on any being, even the lowest, for the sake of gratifying his lower passions. Life must be taken at times to support higher life, and for scientific investigation; even then no unnecessary suffering should be caused. Destructive insects, venomous reptiles, and dangerous wild beasts must be destroyed to protect humanity, but there should be no cruel or long drawn-out tortures. Life is sacred and should never be destroyed without sufficient cause, and then with the least pain. Ruthless destruction of even the lower forms of life has a bad effect on the character and disposition of the children. The boy who takes pleasure in pulling a live fly to pieces, or in crushing a worm under his heel, will soon become hardened enough to destroy the nest and kill the young birds simply to see the anguish of their parents. Cruelty to domestic animals whose lives are given to his service is sure to follow, and this in turn leads to unkindness towards his human associates. And even if such a boy, or the man into whose likeness he grows, never commits murder, he is reasonably sure to become an Ishmaelite whose hand will be against every man, and every man's hand against him.

Life Is Sacred.

Yes, lead the children to observe the animals in their home neighborhood, to study their habits and to notice how admirably they are fitted to serve the purposes of their creation; but at the same time teach them to be kind to all living creatures. Impress it upon them that gentleness begets gentleness, that kindness leaves no sting, and that mercy towards even the inferior forms of life is an indication of a manly character.

CHAPTER IX.

IMPORTANCE OF LABOR.

The last chapter treated briefly of the manner in which soils are formed, and of vegetable and animal life. A way of approach to the study of the vegetable and animal forms found in the environment of the children was shown. The importance of the work as a preparation for the intelligent study of these forms of life in remoter regions was urged with some degree of earnestness; and it is believed that the subject called for still greater earnestness, as it is unreasonable to expect children to appreciate descriptions of the flora and fauna of regions a thousand miles away, unless they are equipped with sufficient apperceiving knowledge acquired by a study of their own locality. Much less will they be able to trace the influence of the physical features and climate upon vegetation and animal life in those far-away countries, unless they have learned to do so, as far as possible, in the vicinity of their own homes. The study of the home neighborhood is the key to the study of the world; and without this key no one can enter into the enjoyments in store for those possessing it.

Key to Study of the World.

It is time to call the attention of the children to the importance of labor. Plants and animals furnish man with food and clothing, or rather with the materials which by work can be made into food and clothing. The tilling of the soil requires labor, so does the rearing of domestic animals. The trees of the forest require much labor to change them into lumber, and then with the lumber to build houses to protect people from the inclemency of the weather. Cotton, wool, flax, and silk would not answer for clothing were it not for the labor expended upon them. And to prepare the skins of animals and make them into boots and shoes also requires labor. By these and other illustrations the children may be led to see the importance and worth of labor in converting raw materials into usable articles.

Need of Labor.

Man needs not only food and clothing; he also needs a fire with which to cook his food and warm his home in winter. In most parts of this country the fuel consists of wood or coal. The children will know the manner in which wood is obtained and prepared for use, but the method by which coal is procured may be new to many of them, especially if they do not live near a coal mine. An explanation of this will introduce the subject of coal-mining. If the teacher has visited a mine, descended the shaft, seen the miners at work, and has noticed how the coal is brought to the bottom of the shaft and then raised to the surface, he can describe the several processes to the children in a more vivid and interesting manner than he can if his knowledge has been obtained entirely from books or hearsay. And they will be led by his description to see the value of labor, and to think more kindly of the men who work several hundred feet under

Mining.

ground, away from the light of the sun, the beauty of the trees and flowers, and the songs of the birds; and who are all the time in more or less danger of being injured or killed by falling rocks.

From coal-mining the teacher can easily pass to iron-mining, and to an explanation of the processes through which the iron passes from the time it is mined until it is made into stoves, railroad-rails, locomotives, etc. Such an explanation can be made both interesting and profitable; and so can an explanation of the processes through which any mineral passes from leaving the mine until made into such articles as best serve the purposes of man. It is advisable, however, that the work be confined mainly, if not entirely, to the study of such minerals as the children can see mined.

Manufactures. There are few neighborhoods in which some articles are not manufactured. It may be that the children see a shoemaker at work day by day as they pass from their homes to school, to the store, or to the postoffice. They may have noticed the harness-maker, the wagon-maker, or the cooper, each working at his trade. The children know that all of these make things, but they may not know that to make things is to manufacture them, and that those who make them are manufacturers. As soon as they learn that to make and to manufacture are the same, they will be ready to name several lines of manufactures which are carried on near home. The gristmill, brickyard, and stove foundry, if accessible, should be visited.

In teaching the subject of manufactures, teachers in the city have a decided advantage over those in the country. Their pupils can visit factories, shops, and foundries, and so get their ideas at first hand, while in some country neighborhoods, the opportunities to do so are very meager; and it is only by the exercise of considerable skill that the teacher is able to interest the children, especially as their parents buy in the nearest town or city such manufactured articles as they need. This condition of things, however, can be used to advantage in leading to the subject of commerce.

While many people are engaged in manufacturing various articles, a larger number are engaged in other occupations. In this country more people are employed in farming, or agriculture, than in any other business. To farm success-

Commerce.

fully, the farmers need plows, harrows, rollers, harvesters, mowers, etc. These they procure by taking their grain, horses, cattle, and hogs to town and selling them for money, and with the money buying such articles as they need. This buying and selling is trade or commerce, and there is so much of it being carried on that a great many men have to devote all of their time to it. Such men are known by various names, as grain-dealers, cattle-dealers, etc., but the general name, merchants, will answer for all of them, although there is a tendency to confine the term to those who keep in buildings, or "stores," the articles they have for sale. The farmer buys, not only his farming implements, but also the clothing for his family, the furniture for his home, and some of his food even, especially tea, coffee, sugar, spices, foreign fruits, etc. Some of these he may be able to buy of the manufacturers, but as a general thing he buys all of them from the merchants, or tradesmen.

It would be of great value to the children if in company with the teacher they could visit several kinds of stores. Many owners or managers would permit the teacher to bring his class to their stores, and would cheerfully devote half an hour to enlightening the pupils as to the places and manner in which his goods were manufactured, and how they reached his store. If only one of these establishments is visited, let it be a grocery store. This will afford excellent opportunities, not only for studying commerce, but also for studying many fruits and other products from foreign countries, and from remote parts of our own.

Good Roads. The children by this time ought to be able to see that good roads are helpful in the carrying on of commerce. Good wagon roads leading from various parts of the country to the towns are essential. If the roads are poor, there may be several weeks at a time in which the farmer cannot take the products of his farm to the maket in town. He is mud-bound. If any of his family is taken sick, it is with great difficulty that he obtains a physician from town. The inconvenience of poor roads and the advantages of good ones ought to be impressed upon the pupils. Some of them will soon be the influential men and women of their communities, and these simple lessons may, in the near future, lead to great improvement in our country roads and village streets.

Railroads. The importance of railroads should also be pointed out, and the manner of their construction explained. The children will understand the subject well enough to pay the teacher for his trouble. All people are interested in railroads, or ought to be. They are a great convenience, as even the children can readily see. By means of such roads they are able to visit friends at a distance in a comparatively short time, protected from storms and cold. Or if they do not care to go themselves, they can send a letter and it will speedily reach their expectant friends. Every town of considerable size has at least one railroad. The pupils have seen the trains standing at the station or moving on the track. They know that one engine can haul as much coal, lumber, grain, etc., to the distant city in one day, as many horses can in a week; and that many kinds of articles would spoil before reaching their destination if they had to be taken in wagons. They admire the engine, and wonder how it is able to haul such heavily-laden trains. A few general facts about its construction, and about the power of steam may be given with profit, if the teacher understands the subjects, and it is reasonable to suppose he does. If he does not wish to do this, or feels unequal to the task, there is something he can do, and ought to do, and that is to point out to the children the impropriety of loitering around the station when they have no business there, or of defacing or injuring the building or its furniture. The children should be made to feel that to scribble on the walls or seats is not only vulgar, but it is positively wrong.

Postoffice. The postoffice also is an important aid to commerce. The children probably have not thought of it in this connection. To them it has only been a means of communication between separated friends. When father or mother, brother or sister, is away from home, they go to the postoffice frequently, hoping to get a letter from the absent one. They are glad when it comes, read it over and over, and hasten to answer it, as they know their friend is anxious to hear from them.

The social value of the postoffice the children will readily appreciate, but its commercial importance may not be evident. The teacher can make it clear, however, by a few simple talks, and while doing so he can impart some useful knowledge in regard to postal matters and letter writing. The size and quality of the paper used, the proper manner to begin and end a letter, how to fold it, and the appropriate kind of envelope, are proper subjects of instruction; so is the place for the address and stamp.

Good taste should be observed in all these matters. Possibly the children will now learn for the first time that the value of the stamp depends on the weight of the letter, and on whether it is addressed to some point in our own country, or to a foreign country. It will not come amiss to follow the letter, in imagination, from the time it is "dropped" into the postoffice until it reaches the person for whom it is intended.

Telegraph and Telephone. The great importance of the telegraph in connection with commerce, as well as in social life, should be made the theme for a few lessons. If the merchant desires to have his goods shipped at once from some distant city, he telegraphs for them, and by so doing receives them sooner than he would had he written, as the wire carries his order faster than the train could. By getting his goods in less time than he would had he sent the order by mail, he may sell them at a larger profit, and also be able to retain the customers who might have purchased elsewhere had he not telegraphed his order. The convenience of the telegraph in sending messages to friends at a distance, especially in cases of sickness or death, will be apparent to the children. It is not necessary, or perhaps advisable, for the teacher to enter into the details of telegraphy, yet it is reasonable to expect him to show its importance and explain the general principles on which the telegraph is constructed and operated.

The telephone is in such common use that it is not thought necessary to urge upon the teacher the duty of explaining its principles to the children. So many of them use it, that in all probability they will ask for the explanation, and he cannot well refuse their request.

Electric Light. The discussion of the telegraph and telephone will lead to the subjects of electric lights and electric street cars. They have much in common. The poles and wires are found in connection with all, while the power that does the work is invisible. A simple explanation of the battery will remove in a measure the mystery which enshrouds the telegraph and telephone, and a visit to the power house will reveal how the force which lights our streets and homes and propels the street cars, is generated. The teacher ought to arrange for such a visit. If it is properly planned and improved, the children will be impressed more than ever with the importance of labor.

Preparation for Book Study. It is hoped that the suggestions made in these articles on the beginnings of geography will remind the teacher of other subjects which he can profitably teach at this time. No better work can be done for the children than to lead them to observe and study the forces that are operating all around them, both in the physical and social worlds. A knowledge of these will put meaning into the operations of like forces in regions lying out of their field of vision, and will render the study of the text-book interesting and profitable.

To make the transition from oral work to the study of the text easy and gradual, brief descriptions of places and industries should be read occasionally in the class. Pictures also can be used to advantage; and simple sketch maps of known localities placed on the board in the presence of the class, will be helpful. To find materials for such maps, short excursions by teacher and pupils are necessary. The places and objects visited should be discussed afterwards in the schoolroom, before being represented by either maps or pictures. When the regions made accessible by these actual excursions are studied, the children will be ready to go with the teacher on imaginary excursions, and not only be prepared to read intelligently the map showing their route, but what is better, they will be able to make the map.

CHAPTER X.

MAP REPRESENTATION.

The following suggestions may assist the teacher in introducing the subject of map drawing:

"We are now ready to make a map of the school room, but neither your slates nor the blackboard is large enough to make the map as large as the floor. So you must let a short line on the board stand for a longer one on the floor, and a still shorter one on your slates must answer for the same line. We can let an inch, or even less, on the board stand for a foot, a yard, a rod, or a line of any length we please; this is making the map on a scale.

Lesson in Map-drawing. "Charles has measured the north side of the room and found that it is thirty-three feet. You may all pass to the board on the north side, and as high as you can reach conveniently draw a horizontal line thirty-three inches long; this is making the map on the scale of an inch to a foot. Alice has measured the west side and found that it is twenty-five feet, so the line that represents it on the board must be how long? It must be what kind of a line? With which end of the horizontal line must it be connected? Remember that we learned some time ago that when facing north, east is on our right and west on our left; hence this line must be connected with which end of the horizontal line? 'The left hand end.' That is right Mary. Since the east side is equal in length to the west side, and the south side to the north, we need not measure those, but may at once draw the lines which represent them. What is the name of the figure we now have on the board? This rectangle stands for what?" etc., etc.

The teacher should have the children point to the side of the room for which each of the lines stands, and the corner which is represented by each of the angles. Let him keep at this drill work until his pupils answer readily and correctly, but not after their answers and general demeanor show that they are weary of the work.

The position of the door, the teacher's desk, the rows of desks, the desks in each row, and everything else that appears on the map should be determined by actual measurement by the class; but nothing should appear that does not touch the floor; hence, neither the windows nor the pictures that hang on the wall can be shown.

This lesson affords an excellent opportunity for reviewing the preceding work. And as the requirements of good teaching demand in each lesson the taxing of the power previously acquired by the pupils, it is desirable that this opportunity for review should be improved.

All objects represented on the map of the school room were where the pupil could see them as he was making the map, hence the work called into activity only his observing powers and his judgment. But in making a map of the school yard and vicinity he will find it necessary to use his memory and imagination in addition to the powers already named, as

Map of the School Yard and Vicinity.

he cannot from his position at the blackboard or desk see all of the objects which he is trying to represent.

Before beginning the map the children should measure the yard and find out its length and width. When this is done they should go to the board and make the map, as the teacher dictates somewhat after the following fashion:

"Place your rulers horizontally against the blackboard and as high as you can reach conveniently. Ready! Draw a line as many inches long as the north side of the yard is rods in length. This is making the map on what scale? On what scale did we make the map of the school room? From the left hand end of this line draw a vertical line as many inches long as the yard is rods wide. This line stands for which side of the yard? How long is it? What is the length of the line that stands for the north side of the yard? The angle formed on the board stands for which corner of the yard? You may now draw a line to represent the east end of the yard. This is parallel to what line? Complete the diagram by drawing the line that stands for the south side. What is the name of this figure? For what does it stand?" etc.

Let the school house be shown in its proper place, and on the same scale as the yard. The position of the gates, coal house, well, and flower beds should be shown by actual measurement by the pupils. If there are no flower beds in the yard, it is the teacher's fault—that is, it is some teacher's fault. If teachers only appreciated the wonderful influence of the beautiful upon children, there would be fewer such ill-kept and disgraceful looking yards as are now found in the towns and villages of the state. Let the teacher cultivate them, and encourage the children to cultivate them, both in the school yard and at their homes. By doing so he will prove himself a public benefactor.

The pupils may next show the streets that border on the school yard, the principal business streets, the streets on which they travel in going from the school house to the postoffice, from the school house to the park, from their homes to the school house, etc. Let the position of the court house, principal churches, and other public buildings be indicated. If there is a bridge in town, let its position be shown, and the direction of the stream which it spans. The railroad station should be carefully located, and the position of the railroads which pass through the village or immediate neighborhood should be plainly shown.

In mapping the streets the scale must be changed, else the map will be too large for the board. It may be necessary to let a quarter of an inch, or even less, stand for a rod. But as the children have mastered the idea of scale fairly well by this time, they will experience no serious trouble from the change.

The teacher should not enter into details too much, and attempt to show too many places on the map. Many err right here. They insist on showing every street and alley in the town, and in some cities even the ward boundaries are shown. This is a mistake due, no doubt, to the fact that they do not appreciate the real purpose of the work. They make the acquisition of facts the chief end, whereas the true end is the cultivation of the children's mental powers; the acquisition of facts being simply a means to an end.

Geography of the Home Neighborhood Before proceeding further in map representation, there should be a study of the land and water forms in the home neighborhood. How shall this be done? The most effective way is to visit the spot under consideration. No verbal description can equal this, nor convey to the mind of the learner so clear a concept of the reality. Unfortunately, however, many teachers find it very inconvenient, if not impossible, to study nature face to face. They have two or three classes in their rooms, and feel that they cannot go out into the fields and woods and down by the "babbling brook" with one class, and leave the others uncared for. Yet, notwithstanding the obstacles in the way, the earnest teacher will manage to take his pupils where they can see the several geographical forms as they are in nature, and so be able later to form approximately correct concepts of like forms which lie beyond their field of observation. He will leave the higher classes under the care of the principal some pleasant afternoon, or give them a half holiday; or possibly he will take his entire school with him and manage to interest all for an hour or more. If the fates decide that he cannot follow any of these plans, he can, and will, take his class on a picnic excursion some Saturday afternoon. Having previously selected the place with reference to the work in hand, and carefully arranged his plans, he can so mingle work with play that the work itself becomes play to the children. And while they go home feeling that they have had a pleasant afternoon, they will carry with them a rich store of geographical concepts to which the teacher can appeal successfully thereafter.

CHAPTER XI.

ANALYSIS AND SYNTHESIS.

In previous chapters attention was called to the importance of a full and careful study of the geography of the home neighborhood. It was shown that without the power acquired from such study the text-book will for a time have but little meaning for the beginner; while with this power he is prepared to enter at once upon the intelligent study of the text. He is able to change the symbols into realities; realities, too, that are familiar, since he has seen their counterparts. The city ceases to be a speck and becomes an orderly arrangement of streets, parks, churches, schools, factories, stores, residences, etc.

Value of Home Geography.

And everywhere are seen people engaged in business, or in pursuit of pleasure. The oceans and lakes are no longer patches of blue or green scattered over the surface of the maps, but are large bodies of water whose waves beat upon the sandy beach or thunder against the rocky cliffs. So with all other symbols which on the map represent forms of land or water; they are transformed by the imagination into the realities which they are intended to represent. The ability to work this transformation — to change the symbol into the thing symbolized, the shadow into the substance, must be acquired by the children before they can use books or maps to advantage. And the only way in which it can be acquired is by a careful study of the home neighborhood, which in reality is the world in miniature.

Adherence to scientific methods would demand that when the pupils have studied all of the world that lies within their field of vision, that is, all of it which they can see and examine for themselves, they should pass to the consideration of the earth as a whole. It is no more difficult for them to see the earth as a unit, through the globe, than it is to see the home state, or even the county, through the ordinary flat map. Whenever they undertake the study of any region lying beyond their field of vision, the knowledge they obtain is mediated knowledge, and it is believed that such knowledge is more likely to be correct when obtained through the medium of the globe than when gained through that of the flat map. They are familiar with many objects which resemble the earth in shape, but with none that resemble the county or state. Consequently they can form an approximately correct mental picture of the earth as a whole more easily than they can of either of the others. It may be urged that the earth is too large for the children to image correctly. True, so it is; and so is the state. The only question here is, in the imaging of which can they approach nearest the truth?

Begin With the Earth as a Whole.

There is another reason for beginning the study of the earth as a whole at this point, and that is that the analytic method will enable the pupils to see the relation of their home to the world long before

Another Reason. they can perceive it if they continue the synthetic method. To work from the home outwards is well, if not carried too far. To go from the known to the unknown is commendable when confined to proper limits. These, like other general statements, need modifying at times. It is wicked to keep children ignorant of the relation just named, until they can obtain the knowledge by aimlessly groping their way from home to the confines of the earth. A wiser course is to begin the study of the earth as a whole as soon as the home geography is completed, follow the analytic plan of study until the relation of the home to the world is seen, and then change to the synthetic mode, and work outwards. This will enable the learners to work intelligently and hopefully, as they see the end to be accomplished.

While the writer fully believes in the reasonableness of the course indicated above, yet in deference to the views of many thoughtful teachers, he is willing to postpone the study of the earth as a whole until the pupils have learned something of the geography of the home township, county, and state. He is the more willing to do so as he is well aware that while they are acquiring this desirable knowledge, their intellectual powers are being so developed that when they do begin the study of the earth as a whole, they will be able to take hold of the work with a firmer grip. So without discussing farther the relative merits of the two modes of procedure, providing the synthetic is not followed at this time beyond the study of the state, he will confine himself to a brief statement of what may be done along this line, and how it may be done to the best advantage.

We will suppose that the children have done the work indicated in the preceding pages. They have made as many excursions for the purpose of study, as the teacher thought feasible, and have studied as much of the neighborhood as circumstances would permit. Not only have the various forms of land and water been carefully noted, but the agencies by which they were fashioned have also received attention. And they have been observed busily at work changing and modifying present forms.

Method of Procedure.

The plants and animals have been studied with a view of determining their habits and usefulness. The leading occupations carried on in the neighborhood were investigated. The convenience and usefulness of railroads, good wagon roads, the post-office, telegraph, telephone, etc., were dwelt upon, and now the children have reached the point where they must enter upon the study of regions which lie beyond the limits of their geographical experience. There is no better way of doing so than for the teacher to lead them on imaginary excursions. In the study of the township and county no books will be necessary. The teacher has probably seen the localities to which he is going to lead his pupils, and is therefore able to describe them both accurately and vividly. If so, the children will get more from his description than they would from a book, as the spoken word is more familiar to them than the written one. Furthermore, back of the spoken word is a living soul, while back of the written one is only dead matter.

Use of Maps. At this stage of the work, and not before, it will be necessary to begin using maps. The study of maps should have no place in the lower grades, but the making o maps should receive considerable attention. If young children are set to studying and copying maps, there is danger that the maps will be the end and aim of their study. Instead of proving a help the maps will be a hindrance, as they will come between the children and the facts to be mastered in such a manner as to hide them from view.

On the other hand, when the pupils have studied a locality, they should be called upon to make a map of it, as soon thereafter as is convenient. Knowing that they will be called upon to do so, causes them to observe forms and relative positions more closely than they otherwise would. They know that the marks or symbols which they place on the map stand for things which actually exist, things which they have seen. The true relation of the map to what it represents will be understood. So that when, later on, it becomes necessary for the pupils to study purchased maps, they will know that the symbols found on these, too, stand for real things. The movement should be from the object to the symbol, not from the symbol to the object.

Township and County. In teaching the geography of the township, county, and state, wall maps which show only such places as the children are to learn, are to be preferred. Maps of the county and state are easily obtained, but it may be necessary for the teacher to place on the blackboard or on good strong paper a map of the township. By the aid of such a map and of vivid descriptions he can lead his pupils to see places and things approximately as he sees them.

The localities studied on these imaginary excursions will at first be comparatively near home, and will therefore have many features that closely resemble those found in the places actually visited, and so will be readily imaged by the pupils. The readiness and accuracy acquired in imaging regions near home will enable them later to form correct mental pictures or images of more remote ones. The growth of the image making power will keep pace with the demands upon it. And when the study of the state has been reached, it will be found that the pupils have acquired such a stock of geographical ideas, and such power of assorting and forming these ideas into new combinations, that with the aid of the text-book and map they will be able to enter upon the work intelligently.

The State. In teaching the geography of the state, the teacher should not attempt too much. This caution is in order at all stages of the children's progress, but more especially at this early stage. Some teachers act as if they believed that the more places their pupils can name and locate, the greater is their knowledge of geography. This is not necessarily true. They may know many places, and yet know but little of the essentials of geography. What should receive most attention are the underlying principles of the science, the foundation facts on which it rests. These, with the relations which bind them into an orderly and philosophic whole, should be uppermost in the teacher's thoughts

For while it is not advisable to dwell on the philosophy of geography with these beginners, the simple facts which are taught them should be so presented that they may be able to see later on that there is a philosophy running through and permeating the study.

The Imaginary Excursion. The surface of the state should be taught clearly and with a fair degree of fullness. The imaginary excursion enables the teacher to do this in a very pleasing manner. He pilots his pupils up or down the river, as he chooses, ascends its main tributaries, carefully noticing the position and direction of the divides that separate the basin of this river from those of the neighboring streams. While doing so, he calls attention to the length and width of the river-basin. the side of the river on which the greater part of it lies, the character of the soil and the nature of the crops. If there is a large city on the river, he points out the relation of its location and growth to the products of the surrounding region. agricultural, mineral, etc. It may be that the growth, or even the very existence of the city is due to the falls near by, affording it great manufacturing facilities which are utilized in converting the forests on the hillside and the minerals in the earth into useful articles.

Or possibly the city is at the mouth of the river and has commerce for its leading business. If so, its importance as a receiving and distributing center for the country back of it is noted. In every instance the relation of the city to its surroundings is pointed out, as relation is the life principle of geography.

By the time the teacher has led his pupils over the principal river-basins of the state in the manner indicated, they will know of its surface, drainage, climate, crops, minerals, manufactures, commerce, and the location of the more important cities. Not many cities should be taken; eight or ten is a sufficient number, but those should be studied quite fully, considering the advancement of the children.

If the teacher is thoughtful and has carefully planned his work, his pupils, in accompanying him on these imaginary excursions, will not only learn the facts just named, but they will also learn what part of the state is wooded and what part prairie. They will locate the principal forests, recognize the most valuable trees, and know for what they are chiefly used. They will learn of the wild animals found in the state and where most of them are found, and will be able to tell in which parts the most domestic animals are reared, and the kind. But better than all the rest, they will know much about the people, both in city and country, their occupations, homes, schools, and general intelligence.

Railroads are at present the principal routes of inland commerce, consequently a few of the more important ones should be studied. This can be done to advantage by going on imaginary railroad excursions from home to the chief cities already learned, as far as circumstances will permit, and from one large city to another. The canals can be studied in the same manner as the railroads. If the teacher has this part of the work in mind when he selects the cities to be learned, the learning of the canals and railroads will require but little time and labor, and they will serve as bonds to hold the cities in their proper place in the mental picture which the pupils are forming.

The Earth as a Whole. Having learned the main facts concerning their home township, county, and state, the pupils should be led to the consideration of the earth as a whole. This topic has been postponed thus far out of deference to the views of some good teachers and, perhaps, the wishes of the parents, but it cannot be put off longer without loss to the learners.

The first fact to be noted should be the shape. To teach this well a globe is essential, the larger the better. By having two or three of different sizes to look at and examine, the children will be saved from thinking that any one of them represents the earth in bulk. To give even an approximately correct notion of the size of the earth requires great skill on the part of the teacher. But the greater the difficulty, the greater should be his ambition to succeed, providing the topic is a proper one. Perhaps the best he can do is to get his pupils to see that it is very, very large; so large that it would take a railroad train six weeks to travel around it, if it were to go as fast as the one on the home road does and make no stops for any purpose whatever.

Other illustrations will suggest themselves and should be used freely, providing they are illustrations that illustrate. The point to be emphasized, however, is the shape, and the globe shows this, as the oblateness of the earth is too insignificant to be introduced at this time.

Motions of the Earth and their Consequences.
The teacher may feel that he ought next to teach the rotation of the earth on its axis, and the relation of the movement to the phenomena of day and night. This is not a difficult topic. It can be taught successfully, if the children know the shape of the earth, and understand that the sun does not revolve around it daily. The revolution of the earth around the sun can be made sufficiently clear to pay for the necessary effort. It is doubtful, however, if the results of the revolution can be made clear. Granted that it is desirable the pupils should understand the cause of the change of seasons, but can these immature minds be made to understand it, even by the utmost efforts of the teacher? The intellectual ability of the pupils to deal with the subject must not be ignored. It is worthy of even greater consideration than is the knowledge to be imparted. The teacher can well afford to defer a full treatment of this topic until his pupils are prepared for it, being satisfied now with a general outline. This course may bring upon him the censure of some thoughtless critic into whose theories of education the capacity of the learners does not enter as a factor. But censure is harmless, if as in this case, it is not merited.

Comparison of Land and Water Surfaces.
The part of the globe that represents land should be compared in size with the part that represents water. The two great masses of land should be compared in regard to shape, size, and direction of greatest length. The oceans should be named, and located with reference to the great

masses of land. It will be well to have a wall map of the world hung where the children can see it, so they can compare the shape and size of the bodies of land and water on the globe with the shape and size of the same bodies on the map. Comparison should have a prominent place in the teaching of geography.

Comparison of Hemispheres and Continents. The terms, eastern hemisphere, western hemisphere, new world, and old world, should be introduced, and the reason for the names given. When this is done the two great masses of land may be divided into continents. The number of continents depends on the definition of the term. It is convenient to regard them as six, including Australia. They can be arranged in three pairs, each pair consisting of a northern and a southern continent. A pupil of average ability will see by the map that several of the continents resemble each other in many respects while they differ in others. All differences and resemblances should be carefully noted.

The continents may also be arranged into groups of threes—three in the northern hemisphere and three in the southern, the two groups being almost separated by a great depression which is filled with water. In each group the continents are wide at the north and grow narrow towards the south.

This comparison of the continents is a fruitful theme. By it the pupils will be able to fix the relative position, shape, comparative size, character of the coastline, and many other facts which it is essential they should know. The teacher should be in no hurry to leave it; but when he does leave it, it should be to study the home continent more fully.

Analysis and Synthesis. 103

Reviews. The facts already learned about the home continent should be reviewed before teaching new ones. In fact there should be brief reviews daily. Two or three judicious questions each day will keep the matter stirred up in the consciousness of the children. This is important. It is difficult to see how there can be any advancement in true learning without this. With this knowledge as a working capital the pupils are prepared to note the irregularities of the coastline, name and locate the principal projections and indentations, the great mountain systems, plateaus and plains, and the chief lake and river systems. Only a few of the more important facts should be considered; enough to enable the children to form a mental picture showing the principal features in their proper relations.

The teacher may be tempted to tarry here and teach much about climate, vegetation, animals, minerals, and the people with their various industries and interests. The wisdom of doing so is doubtful, to say the least. The purpose now should be to get back to the home, marking the road so plainly, that when the pupils start out on their conquest of the world, they will know the route to be traveled, and see the relation of each day's journey to the starting point, and to the entire work to be accomplished.

Home Continent. On the home continent are situated several countries. Their boundaries are determined by men, and are arbitrarily fixed, often without reference to natural limitations. Consequently the political and natural divisions rarely coincide. So the best we can now do is to teach the names and positions of the countries with a few general facts about each; the number of facts to be determined by the relative importance of the countries. The home country should of course receive most attention. Even this should not be studied too much in detail. Its position on the continent, its shape, size, climate, and in a general way its varied productions should be noted. The natural features should be reviewed. And it may be well to teach a few of the principal cities and railroads. The position of the home state must be fixed, and its relation to the country as a whole noticed. When this is done, the pupils will be prepared to move intelligently "from the home outwards;" to go "from the known to the unknown;" and there is no better or more interesting way of doing so than the imaginary excursion.

CHAPTER XII.

VALUE OF MAPS AND PICTURES IN TEACHING GEOGRAPHY.

Pictures. Next to the object itself, its picture excites the greatest interest, and conveys to the mind the clearest and most accurate impressions. Besides appealing to the eye, which is valuable, if it is a good picture it will also awaken the innate love of beauty with which we must credit the average child. The wealth of illustrations found in our text-books shows that pictures are regarded by teachers, generally, as important aids in education. Were it not so, publishers would not go to the trouble and expense of inserting them, as they are not in the habit of expending their money upon that which is naught. The illustrations in our leading geographies are pleasing to the eye, and, in the main, truthfully represent the objects for which they stand. Yet there are teachers who either entirely ignore these pictures, or refer to them as pictures, simply, without connecting them with the objects. This is a misfortune, as the illustrations are sometimes as valuable as the text.

Geographical pictures are abundant, and can be obtained at little expense. The teacher should have on his desk, or on the reference table, copies of several geographies. This will give him quite a variety of pictures. In addition, he can obtain beautifully and profusely illustrated pamphlets from the agents of the principal railroads by sending a few postage stamps. Several such lie on my table as I am writing. In one, issued by the New York Central and Hudson River railroad, I find a view of Niagara Falls a glance at which will give a child more knowledge than will ten or fifteen minutes spent in studying a verbal description. Here, also, are views of the Hudson river at several points, showing the Palisades, the Highlands, and several other famous localities. The valley of the Mohawk, the Erie canal, and the lakes of northwestern New York offered many beautiful scenes, which the artist has kindly placed here for my benefit. The West, not to be out-done, has sent me "Indian Land and Wonder Land," all the way from St. Paul, for six cents. An artist friend of mine tells me that the illustrations in this volume are the finest he has ever seen in such books; in fact, that they are gems of art. The first one to which I turn satisfies me that my friend's judgment is good. It is a picture of Lake Chelan, in Washington. On one side of the lake are seen a few huts, several fields and a number of men busily at work, showing that settlers have been attracted by the beauty of the location and the fertility of the soil. On the other side are the mountains, whose shadows are thrown far out upon the water. This is but one of many scenes depicted by the artist. The Bad Lands of North Dakota, the Yellowstone Park and Alaska, are represented by barren wastes, towering mountains, placid lakes, steaming geysers and foaming cataracts; all of which can be made very serviceable in the geography class. Railroads are not the only corporations that advertise their business in such artistic forms; publishers and manufacturers also advertise by means of illustrated pamphlets, which they are glad to send to teachers who ask for them and are willing to pay the postage.

Stereoscopic views can also be used to advantage. They are not very expensive, do not occupy much space, and can easily be carried from place to place. This last is quite an item in their favor, as the teacher's health may cause him frequently to move from district to district. I use this class of pictures somewhat freely in my teaching, and I **Stereoscopic Views.** should use them more freely still were I teaching boys and girls. My experience leads me to value them highly. I find that if a pupil has looked at a picture of the locality, or object, he will get much more from reading a description of it, or from listening to a description by the teacher, than he will if he has not seen the picture. He sees the object through the picture, and so has something tangible to which to apply the words. This is true of any good picture.

Pictures not only aid the pupils in getting clear geographical notions of those parts of the world which they have not seen; they are of great assistance to the teacher as well. They inspire him to paint vivid word-pictures of objects and localities, a gift which every teacher should diligently seek. A bird's-eye view of the basin of Lake Champlain, bordered on the east and west by mountains, whose ramifications extend well down toward the lake, diversifying the surface with sunny slope and shady vale, traversed by sparkling streams, and dotted with cities and villages, many of which are mirrored in the clear waters of the lake, while others are surrounded by well-kept farms, whose sleek herds and fine-wooled flocks show the thrift of the owners, will materially aid the teacher in describing any lake basin. In like manner a good picture of a river basin will help in describing other river basins; and a good representation of a woodland scene will assist in picturing other woodland scenes, etc. I place such high estimate on this power of word-picturing that I have no hesitation in saying: Blessed are the pupils whose teacher has this gift, provided it is seasoned with good sense; otherwise he may be airing his ability in season and out of season, and the gift may do more harm than good.

Word Reading.

Looking at pictures, however, should not be made the main purpose of the recitation. They are valuable, simply, as <u>means to an end</u>, and the end in this instance is the acquisition of geographical knowledge. If the examination of the pictures is made the end, then the pictures are hurtful, and will lead to dissipation of energy and waste of time, as the attention of the pupils is drawn away from the geography, and fixed upon the pictures as pleasing objects merely. This will be especially true of the younger children whose will-power is weak and whose craving is strong for that which is pleasing. Hence the need of caution. Without its exercise a good thing may be rendered worse than useless.

Maps.

Geography is not a study of words and maps. It is a study of the world in which we live, work, and go to school. It calls attention to the various forms of land and water, to all forms of life, but more especially to man and all that concerns him. The atmosphere is studied and its importance shown. In the lower grades these facts, forms, and notions are studied only as they exist in the home neighborhood. Only such forms as lie within his field of vision and such facts as find a response in his own experience, must be presented to the young learner.

Above the primary grades the printed map can be made helpful, but not before. Even here there is danger that it will be looked upon as the real object of study, instead of being regarded as a symbol. I am satisfied that there are many pupils who spend weeks or months on the geography of North America without once thinking of the real continent. Its great extent, snow-capped mountains, broad plateaus, fertile plains, inland seas, majestic rivers, peaceful farms, busy cities, etc., are to them as though they did not exist. Their thoughts are centered on the map that hangs on the wall, or on the smaller one that is in their text-book; but they do not appreciate the true function of either.

There must be a constant effort on the part of the teacher to get his pupils to see that the map is a symbol, and that its purpose is to enable the mind to pass over from the words to the forms of real land and water, which are described by the words. When the map is regarded in this light it is a valuable aid in teaching, as it suggests a picture, and geography may justly be regarded as a series of related pictures.

All maps are helpful when pupils are taught their purpose, but the raised map is the most helpful. It excels the flat map, not only because it shows the vertical configuration, which the other does not, but because it shows the contour even better, as it suggests that it is determined by the character of the surface. It is true that some object to the raised map because altitudes must be exaggerated and out of all due proportion to horizontal distances, in order that the vertical forms may be appreciable. Admitting that there must be exaggeration, yet is it not a fact that notwithstanding this fault the raised map enables the children to form a truer mental picture of the surface than does the flat map? I believe this to be true; consequently I favor its use, and urge all teachers of geography to get a good supply of the best raised maps. With all their imperfections they will lead the children nearer the truth than will the flat maps; and the truth is the desired goal.

Next in usefulness to the raised maps are the physical maps, of which Guyot's are among the best. These maps show different altitudes by means of different colors. They enable the pupils to see the location and directions of the mountain system, the position and comparative size of the plains and plateaus, and the position of the deserts with the reasons therefor. Notwithstanding that these maps give an opportunity for using but one sense while the raised maps invite the use of two, yet they are desirable helps, and by an intelligent use of them good results may be obtained.

It is hardly necessary to spend much time in discussing the helpfulness of a map that does not show relief by some device. There are but few such, and they are rapidly passing away; and let them go. May the school houses that now know them, soon know them no more. In the hands of skillful teachers they did some good; when used by others they often did harm. They are far inferior to the physical maps, yet cost nearly as much. There is, therefore, no reason other than the oratory of the agent for buying any more of them.

Sand Modeling. Of all the devices used in aiding the imagination to picture remote geographical forms, sand modeling is the most serviceable. By its aid forms of contour and relief can be shown as readily as by the raised maps, and they can be altered easily. The pupils can be made to see the continent, island, or peninsula growing under the hands of the teacher. This will be more likely to inspire them to put forth effort than will the looking at a map which is unalterably fashioned, when it is brought before them. By seeing things made, children will learn to make them more readily than they will by hearing how they are made, as they will be more interested. Here the various forms are fashioned gradually as they were in nature, and the nearer nature we can keep the device, the better. This is why the molded form suggests the real land or water more readily than does the map.

The sand is also very helpful in enabling the children to understand how the surface of the earth has been changed, and is constantly being changed by the action of the air, frost, water, etc. This cannot be shown by the aid of maps, yet the learners' attention should be called to it, so they may be led to observe the work of the great artists who are constantly carving and chiseling, wearing down the sharp ridges and filling up the unsightly hollows that mar the appearance of our earth-home.

Another advantage that sand modeling has, is that it can be used in all the grades from the primary to the high school. It is especially valuable in teaching physical geography; but as that is the basis and most important part of all geography, it is reasonable to suppose that it is taught throughout the entire course.

Objections are sometimes made to sand modeling because of the exaggeration necessary, and because of the danger that the children will use the forms shown in the sand as the standards with which to measure like forms in nature. What has been said in defense of raised maps will apply here equally as well. And it is safe to say that no evil will result from the use of the sand or raised maps if the teacher exercises due care.

The Imaginary Excursion
and its Place in Teaching Geography.

CHAPTER XIII.

A TRIP DOWN THE HUDSON RIVER.

Proper Interpretation of Map.
To read a map correctly is quite an achievement. The ability to do so is a prerequisite to the intelligent study of all regions which cannot be visited by the learner. The accuracy of the mental picture formed depends on the power to interpret geographical symbols, and the value of the description is determined by the character of the mental picture. This is true in the study of most subjects, but more especially in that of geography, history, literature, and reading. No matter how eloquent the description, or how faultless and beautiful the composition, it has but little educative value unless there is in the mind of the reciter a fitting series of images. His belief that this is so, accounts for the frequency with which the writer has urged the importance of aiding the children to make correct mental pictures. For while it is not all of geography to be able to do so, it is essential to any marked success in pursuing the subject.

The pupils with whom we have labored thus far, ought to possess by this time sufficient power to interpret the map correctly, and to appreciate the description. Much strength was acquired along these lines by accompanying the teacher on the imaginary excursions which he made to different parts of the home township, county, and state. Since then he has led them on longer trips. Together they have sailed on longer rivers, climbed higher mountains, and traveled over more extensive plateaus. They have witnessed the buying and selling which are constantly taking place at the great centers of commerce, and have broadened their knowledge by tracing the relation between the agricultural products offered for sale, and the climatic and other conditions of the countries which produced them.

Value of Excursions.

Manufacturing centers have also arrested their attention, and they have stopped long enough to examine the lines of goods manufactured, and to discover, if possible, the causes which led to their being manufactured at those particular places. The location may be due to abundant water power, nearness to inexhaustible supplies of fuel, and convenience to market; while the nature of the goods may be conditioned by the character of the raw materials most accessible.

In the fields have been noticed men at work planting, cultivating, and harvesting. The manner of caring for some of the leading crops has been noticed, as well as the dependence of certain products upon the climate and soil. On the higher plateaus, where agriculture is unprofitable, cattle and sheep were seen in great numbers, being fattened for the market. And on plain, plateau, and mountain, men were observed digging into the bowels of the earth and bringing forth the minerals which add so much to the wealth of the nation. Having seen these things with their own eyes or those of the teacher, they are prepared to make longer journeys in their own country and in foreign parts. The first lengthy excursion will be a trip down the Hudson river.

The Appalachian mountains are peculiar in their formation. In places they consist of parallel ranges separated by fertile valleys. In others they are wide plateaus, dotted here and there by isolated peaks. While in still other places they form groups of steep and well-rounded mountains, in many of whose deep valleys are found bright and silvery lakes.

The Adirondack mountains constitute one of these groups. A few years ago they were but little known; now they are among the most popular resorts in America. A few hours' ride takes one away from the hurry and bustle and noise of the metropolis of the western hemisphere to the solitude of the wilderness, where he will find strength for his tired body and rest for his fevered brain.

From out the heart of these mountains flows one of the noblest rivers of the world, the lordly Hudson. It is true that it is surpassed in length and volume by the Amazon, Mississippi, and a score of others, and that it cannot boast of the ruined castles of robber-knights of the queenly Rhine. Yet it is doubtful if there is anywhere on earth three hundred and fifty miles of river that surpass the Hudson in the grandeur of its scenery, in the beauty and pathos of its legends, or in the deeds of daring and cruelty that have taken place along its banks. The dusky beau has sent his bark canoe skimming over its waters to meet his peerless one at the secluded

The Hudson River.

trysting place. The Mohawk brave has noiselessly moved in the shadow of its overhanging trees, stalking the enemy of his tribe, as the hunter stalks the deer. And armies of freemen have put to rout the hosts of the oppressor within the sound of its rippling waters.

The Legend of Minne-wa-wa! Among the most famous Indian legends of the Hudson is that of Minne-wa-wa, which is as follows:* "In the valley of the Hudson lived the Mohicans, who were the direct descendents of the Great Spirit. Minne-wa-wa, the pleasant voice, was the mother of their chief, who was called the Evening Star, and who had for his wife Wa-bun An-nung, the Morning Star; their son was named Osseo, Son of the Evening Star.

"Soon after they settled in these pleasant hunting grounds, Osseo and his father, while chasing the red deer among the blue mountains that lie to the west of the sparkling river, were overtaken by Mishe-mokwa, the great naked bear, and destroyed.

"Wa-bun An-nung in her great sorrow wandered away from the village to the east, and was taken by the Puk-wud-jin-inies, the little vanishing men of the woods who are seen as night approaches, and suspended in the eastern sky, where she became the morning star.

"Minne-wa wa, bereft of all kindred, betook herself to the western mountains to grieve in solitude near the spot from whence her loved ones had vanished. As time, that great healer of human woes, somewhat assuaged her grief, her heart beat only for the subjects of her lost son, and her greatest de-

---

*Taken from "Hudson River by Daylight," by permission.

sire was for the welfare of her people; and fearing lest some others of the tribe might be overtaken in the darkness by Mishe-mokwa, she gave to the little swamp-flies, Wah-wah-tay-see, the wee lamps which at night they flash here and there among the bushes, that they might reveal the monster should he be lurking near to devour. But perceiving that the Wah-wah-tay-see would be of service only in the damp hollows which they frequented, and under the shadows of the dense trees and thick bushes, she climbed the mountain, and from the highest peak hung in the western sky the crescent bow of the lost Evening Star, to which she gave light and which became the moon. The Great Spirit, seeing that this was good for her people, and that she of all others held their welfare in highest esteem, changed her into an immortal spirit and gave her the vast mountain for a lodge, in which was the great treasury of storm and sunshine for the region of the Hudson; and he gave her also the dispensing of them for all time. Here she kept Day and Night shut up, letting out only one of them at a time. Monthly she came from her dwelling and hung the crescent new moon in the western sky, over the mountains, and so placed it as to signal to her people that she was about to send out the cooling showers to water the maize-fields and freshen the springs and parched herbage; for if she so hung it that the lower horn of the crescent was elevated sufficiently to hang upon it the bow and quiver of the hunter, then was the hunting at an end for a season and her people were to keep to their lodges and wigwams.

"After hanging forth the signal that all might see it she would stand on the mountain top and shake from the folds of her mantle the drifting rain-clouds, and blow them over the valley with her breath. Sometimes she would weave them out of cobwebs, gossamer, and morning dew, and send them off, flake after flake, to float in the air and give light summer showers. When the people had done that which displeased her she would brew up black thunder storms, and send down drenching rains to swell the streams and sweep everything away, and with them, the thunder of her voice and the lightning flashes from her eyes.

"Thus did Minne-wa-wa become the guardian of the Mohican people, ever ministering to their good, sending the rain to moisten the maize fields, and water the hills that the herbage might grow and keep the game in abundance for them. Monthly she hung up the new moon, and as often cut up the old and scattered the little pieces throughout the heavens, and made of them the little stars whose lamps she lighted nightly."

The Hudson river rises in Henderson lake at the foot of Mount Marcy, the highest peak of the Adirondacks. Its course at first is through a rough and rugged country. It works its way through dark and cavernous glens, whose beetling cliffs shut out the light of the sun. Its roar may be heard echoing among the mountains as it goes fretting and fuming through the rapids, or wrathfully throws itself over the rocky precipices. Occasionally it comes to a small level meadow through which it peacefully flows singing anthems of victory over the difficulties it has overcome, and gathering new strength for the obstacles that may still be in its way.

The river flows nearly south for quite a distance, when it is compelled by the hardness of the rock to take a northeasterly course until it reaches Sandy Hill, when it again turns to the south and continues in that direction to New York bay.

Shortly before reaching Sandy Hill the river is divided by a small rocky island and dashes over a precipice sixty-three feet in height. Both the island and falls are made famous by Cooper in his *Last of the Mohicans*. In the rocky caverns of the island Duncan Heyward and his wards, Alice and Cora Munro, sought safety from their savage foes. It was here that Hawkeye and his two loyal Delawares fought the "Mingoes" until their ammunition was gone and then plunged beneath the seething waters and made their escape. And it was here that David Gamut, inspired by the noise of the cataract, poured out his soul in nasal melody. But alas! neither the singing of David nor the faithfulness of Duncan saved the party from falling into the hands of Magua, the Mingo chief. The village of Glen's Falls marks the spot at present.

With its escape from this wild region, the Hudson enters a more level country and its career for the rest of its course is comparatively peaceful. But while escaping the turbulence of nature, it has witnessed many of the turmoils of men, both savage and civilized. There is hardly a foot of land along its banks from Sandy Hill to New York bay that has not been fought over by armed men; and the same may be said of its principal tributaries. Hurons,

Mohawks, Dutch, English, and Americans have dyed the water with their blood and enriched the soil with their bodies. The war whoop of the Indian, the slogan of the Highlander, and the hearty hurrah of the English and Americans have reverberated from its cliffs; but heedless of them all it has kept on its way, anxious only to find repose in the arms of its great mother, the ocean.

The village of Fort Edward is a short distance below the great bend. The portage from here to Lake George was the most dangerous part of the route to Canada. The village figures conspicuously in the story of Burgoyne's invasion.

Bemis Heights stretch along the west bank of the Hudson from opposite Fort Edward to the Mohawk river. The altitude of these heights is not great at any point, and in some places they are slightly undulating plains. Here were fought two important battles of the Revolutionary War. In one of them Benedict Arnold won considerable renown, although Gates received the official credit. The last of these battles is vividly portrayed by the wife of a Hessian general who was in the service of the English. She gives great credit to the Americans, not only for their bravery in battle, but also for their chivalric conduct in not firing on her servant maids who supplied the wounded British with water from the river. The report is found in *Lossing's Field-Book of the Revolution*, also in *Washington and His Country*. It is well worth reading.

Saratoga Springs. About twelve miles west of the battle field is the village of Saratoga Springs. This is one of the most noted watering-places in America. In the fashionable season its twenty or more hotels, with many private boarding houses, are filled to overflowing with those in quest of health and pleasure. Its thirty-odd medicinal springs are patronized by those who are sick, or think they are; but for most the great attractions are the costly equipages and brilliant costumes that may be witnessed on its streets, and the gay balls and parties that may be attended at night. The natural scenery in the immediate vicinity is not very attractive. The "sights" have been produced by man's labor and skill. August is the best time to visit Saratoga, if one desires to see it in its glory.

Mohawk River. The Mohawk is the chief tributary of the Hudson. It flows through one of the principal depressions of the Appalachians. In colonial times it formed one of the leading routes to Canada, there being but a short and easy portage from its headwaters to those of the Oswego. It was then the dark and bloody river, being guarded by the terrible Mohawks, the bravest and most bloodthirsty of savages. Now it is a helpful servant of civilization. Its rapids and falls are utilized in turning the wheels of industry; its fertile valley is occupied by peaceful farms, thrifty towns, and well-ordered cities. Instead of the war whoop of the savage is heard the whistle of the locomotive and factory, and the site of the "long house" of the aborigines is occupied by the church and the schoolhouse, harbingers of good to man. And as if to seal the conversion of the river to the service of humanity, along its banks extends the Erie canal, a messenger of peace between the seaboard and the interior.

Schnectady. Schenectady is situated on the Mohawk river and the Erie canal. The very name causes the student of history to shudder as he thinks of the stockaded village, the faithless sentinels, and the defenceless inhabitants wakened from their sleep to see, by the light of their burning homes, the ruthless savages standing over them brandishing tomahawks and scalping knives. The horrors of the flight to Albany through the snow, barefoot and scantily clad, was equaled only by the plight of those carried captive to Canada. Happily the town is now an important manufacturing center, and its people are safe from the perils which surrounded their ancestors.

Troy, the largest manufacturing city in the valley of the Hudson, except New York, is situated on the east bank of the river, on an alluvial plain. Its principal manufactures are iron, steel, stoves, railroad cars, etc. Being at the head of navigation, and having a number of railroads, it is also largely engaged in commerce. It is the seat of the Rensselaer Polytechnic Institute, and across the river is the Watervliet national arsenal in which are made large cannon. The tide ascends to Troy.

Albany. Albany, on the west bank, is the capital of the state of New York. It is the eastern terminus of the Erie canal, and is connected with Lake Champlain by the Champlain canal. Several lines of palatial steamboats make regular trips between Albany and New York. In summer they are crowded with pleasure seekers who never tire of the magnificent scenery which the trip reveals to them. The city has several railroads which with its water facilities make it an important commercial center. It is also extensively engaged in commerce.

The Catskill Mountains. We leave the river for a time at Catskill Landing, on the west bank, and reach the mountains by railroad. On arrival we hasten up the rocky valley that we may witness Rip Van Winkle assist his taciturn companion in carrying the heavy keg. If there is a thunder storm we may hear the noise of the balls, as they are rolled by the crew of the Half Moon, long before we reach the amphitheater. A motley crowd they are and freely do they partake of the contents of the keg. But having Rip's sad fate in remembrance, we keep at a safe distance from the flagon. Becoming tired of the play we ascend one of the highest peaks, and sitting by the side of **Leather Stocking**, or **Hawkeye**, as we called him elsewhere, we listen to his marvellous tales of Indian cruelty and the white man's cupidity.

Leaving the past, however, and addressing ourselves to the present, we soon realize that we are in a region rich in scenic grandeur. Mountain peaks, plateaus, and deep-furrowed valleys surround us on all sides. Deep, placid lakes mirror the beauty of their wooded margins, doubling our pleasure. Brisk, refreshing breezes fan our brows, and our ears are delighted by the music of many streams as they go dancing over their rocky beds and between their grassy banks, breaking out into a joyful chorus, loud and deep, as they descend to the plain. The most noted is Kaaterskill creek which plunges down three hundred feet at one leap.

From our point of observation we can see the Adirondacks to the north, with their tops reaching to the heavens. Between lies the valley of the Mohawk in which the river appears like a ribbon of silver.

Turning to the east, the broad valley of the Hudson is spread out before us. Cities and towns dot the landscape. The clouds of smoke which rise above them tell of labor, inventive genius, and proud achievement; while the spires that glisten in the sunlight reveal their hopes when life's labor is ended.

Thrifty and well kept farms may be seen on both sides of the river. The elegant residences and capacious barns show that the tillers of the soil are well repaid for their labor. Groves of hemlock, maple, and oak give variety to the scene; and the herds of cattle, droves of horses, and flocks of sheep in the carefully fenced pastures add life and motion. The Hudson itself flows by at a distance of seven miles from the base of the mountains, carrying on its bosom the commerce of an empire. All sorts of craft, from the plebeian canal boat to the aristocratic steamer equaling in its appointments the palace of a king, ceaselessly plow its waters; while on its banks may still be seen many of the stately mansions built by the Dutch Patroons, alternating with the modern villas of merchant princes. For ninety miles up and down the valley and east to the Green mountains and Berkshire hills, this enchanting panorama is unfolded to our gaze. On the south it is limited by the Highlands.

The view to the west is of a more quiet nature, yet it is not without beauty. The broad, fertile valleys of the Delaware and Susquehanna rivers appear in the distance. Signs of comfort, if not of opulence, are seen in the substantial homes that greet our eyes. Brawling cataracts and rocky crags with their weird and picturesque scenery are absent, but tokens of peace and plenty are everywhere visible.

Returning to the Hudson, we soon reach Kingston, on the west bank. This is the eastern terminus of the Delaware and Hudson canal, which is used chiefly in conveying coal from the anthracite regions of Pennsylvania to New York city.

A short distance farther on is found Poughkeepsie, an important manufacturing center. It is especially noted, however, for its educational institutions, of which Vassar College, for young ladies, is the principal one. The beauty and boldness of the great cantilever bridge spanning the river at this point is worthy of our admiration.

In passing by Newburg, we are pleased with its very attractive appearance. It was here that Washington, at the close of the war, by his energy and prudence saved his country from a possible revolt of the army. Across the river is Fishkill, not far from which were the haunts of Cooper's "Spy."

The Highlands. As we approach the Highlands, we notice that the river is narrowing and the velocity of its current increasing. We enter the enchanted region between two high rocky piles that stand as grim sentinels guarding the passage. The one on the west is Storm King. It looks solemnly across at Breakneck to see if it is attending to its duty.

We have no trouble in convincing these guards that we are friends, who simply wish to view the treasures entrusted to their custody. The scenery on both sides of the river is beautiful, grand, majestic. Rock rises above rock, cliff above cliff, and mountain above mountain. While opening back from the river are many narrow valleys which serve to make the massiveness of the mountains all the more impressive. Several of the mountains look upon the river with lowering brows, as if displeased with it for bringing strangers to gaze upon their secrets. Some of them are even so angry that they run out into the water determined to stop its passage. It moves on, however, and moving with it, we arrive at West Point.

A visit to the "Point" is apt to recall many events in the history of the nation. It was here that Arnold, the traitor, tried to undo all that Arnold, the patriot, had done. The names of Washington, Kosciusko, and Lafayette are associated with the place, and are held in precious remembrance by the young men who are here being educated to follow in their footsteps.

West Point.

The military academy occupies a plateau one hundred and eighty-eight feet above the river. It is generally conceded to be one of the best schools of its kind in the world. The careers of Grant, Sherman, Sheridan, and of many other great captains, bear witness to the excellence of the training received here.

We stay at West Point long enough to visit the cadet barracks, the library, the riding school, battle monument, and "Lover's Walk." We see the cadets on dress parade, and visit mess hall, later, when they are partaking of their evening meal. They are a bright, vigorous, and jolly body of young men, and they take pains to make the visitor's stay among them as pleasant as possible. Unfortunately our time was limited, so we had to say good-bye to these future heroes, and continue on our way.

We leave the Highlands, as we entered them, between two peaks; "for the Dunderburg and Manito stand guard at the south, rearing their heads skyward more than one thousand feet." In passing Stony Point, we think of General Wayne and of his heroic capture of the fort.

It is not difficult at this juncture to turn our thoughts from war to peace, as the river spreads out into the broad, quiet expanse known generally as Haverstraw bay, but whose lower part is locally designated, the Tappan Zee. The shores consist no longer of mountain walls, but in many places, the land slopes down to the water's edge. And while there are occasional cliffs, the prevailing characteristic of the scenery is its tranquility. Every point of vantage is occupied by a beautiful villa whose surroundings are as pleasing as money and skill can make them. Farms stretch back from the river on either side, but the growing crops do not indicate fertility of soil.

The Croton river comes in from the east. From the upper valley of this stream New York city is supplied with water by means of an aqueduct forty miles long.

Tappan is on the west bank of the Hudson. It was here that Major Andre, the British spy, was tried by court martial and executed. He was captured east of the river, just across from the village.

Tarrytown is the next point at which we stop, and here we disembark for the present. The town "is delightfully situated on an elevated plateau, overlooking the wide expanse of the Tappan Zee and the surrounding country for many miles." We do not land, however, simply to see the town; we must visit "Sleepy Hollow," which is quite near. The brook still glides through the little valley murmuring as it did when Ichabod Crane, stretched upon its bank, had such ecstatic visions of the wealth of Baltus Van Tassel and of the plumpness of blooming Katrina, all of which he hoped would soon be his.

Sleepy Hollow.

We wander in the churchyard looking for the grave of the unfortunate Hessian but we find it not. We are shown the very spot by the brook where was found Ichabod's hat and the pumpkin. Who can tell how the pumpkin came to be in that particular place on that awful occasion? As we are trying to solve the mystery, we can almost hear Gunpowder tearing down the road, closely followed by the headless horseman. That pumpkin, so harmless, seemingly, brought ruin and destruction upon the hopes of the schoolmaster. We can readily imagine the feelings of Brom Bones when he heard of his rival's mishap.

Returning to the river once more, we soon come in sight of "Sunnyside" cottage, the home of Washington Irving. Remembering the kind, genial nature of him who made it famous, and acknowledging our indebtedness to him for many an hour of pleasure and profit, we reverently salute it, and look upon it with uncovered heads as it is rapidly hidden from sight by a dense growth of trees and shrubbery.

The Palisades is a name applied to a long perpendicular wall that extends for twenty miles along the west side of the Hudson, terminating on the south at Fort Lee. They form the river edge of a plateau about three-quarters of a mile wide, and from one hundred to five hundred feet high. In many places the face of the cliff presents a columnar appearance. The opposite, low, verdant shore affords a varied and charming picture from the Palisades; while to the south the eye reaches to the metropolis and its crowded bay.

The towns and villages on the east bank follow each other in rapid succession. No sooner do we pass one than we come in sight of another. We do not land at any of them, as we desire to reach the great city by daylight, and the sun will soon sink below the horizon. Spuyten Duyvel creek is passed. The river is well nigh covered with boats of all descriptions. We carefully pick our way among them, and soon are in front of Riverside park. Here we rest on our oars a short time and think of the hero whose tomb is in plain view. The spot is a lovely one. The view up and down the river cannot be surpassed. Long may it be before the American passer-by forgets to look with reverence upon the last resting place of the immortal Grant.

We land at the 22nd Street pier, and are driven to our hotel, feeling well repaid for our trip down the Hudson.

CHAPTER XIV.

A TRIP DOWN THE RHINE RIVER.

The drainage of at least 150 glaciers unite to form the headwaters of the Rhine river. The principal stream rises near Mt. St. Gothard in an icy cave, amidst a mass of rocks. It flows to the northeast for some distance, between frowning walls and in dark gorges, and then forces its way northward through several mountain ranges, until it reaches the peaceful waters of Lake Constance. The descent is so rapid in the upper part of its course that we can not use our canoe. We must carry it the best we can, for were we to launch it, it would soon

The Upper Rhine. be dashed to pieces in the rapids through which the water rushes, swirling and foaming. If we look up the high, narrow valleys that open out on either side, we shall see small huts built under the shelter of the cliffs. These are occupied by hardy mountaineers who derive much of their livelihood from their small flocks of goats. Every patch of grass is utilized. In some instances the herdsman may be seen carrying the goats on his back to some little grassy plateau which they can not reach otherwise. Farther down the valley the mountains recede from the river, and cattle are seen grazing along its banks. Small villages appear, whose inhabitants cultivate a few of the hardier grains and vegetables. The houses are small, rudely built, and scantily furnished. Yet the people are happy, if we may judge by the songs and shouts of merriment which we hear so frequently.

The People. For some distance before reaching Lake Constance the river is on the boundary between Austria and Switzerland, and flows through a great alluvial plain which is supposed to be a filled-up lake basin. The Rhine is now a well-behaved river, and is navigable for small boats and rafts. We embark in our canoe, and as we are borne along by the current we notice thrifty towns, carefully tilled fields, and rich meadow lands. The farms are small, and so the tillers are able to live in hamlets, or small villages. Isolated farm houses are rare. A village consists of a row of one-story houses on one or both sides of the road. It is rarely that one of them has more than one room, although a few of the more pretentious have two. But whether one room or two, they usually shelter the domestic animals, as well as their owners. The farming implements are very primitive. The spade, the hoe, the sickle, the scythe, and the hand-rake are still the main reliance. The horse and ox do but little of the work; woman is the more common beast of burden. She carries the fertilizer from before the house and spreads it on the land, often with her hands. In the autumn she carries the hay and grain from the field and stores them away in the rude loft overhead, or stacks them near by. Her husband, brother, or employer may help her get the burden on her back, but he will not carry it, as he considers it unmanly to do so. Everywhere in Europe the lot of the peasant woman is a hard one. Her labors are many and arduous, and her joys are few.

Lake Constance. Austria, Germany, and Switzerland border on Lake Constance. The mountains in many places come close to the lake, and their shadows are cast upon its waters by the setting sun. Quaint-looking towns, surrounded by large apple orchards, dot its shores. Their narrow, crooked streets, peculiar architecture, and "mild flavor of decay" attract our attention, and lead us to land at one of them. We are cordially welcomed by guides, hotel-keepers, merchants, and others who hope to derive profit from our presence.

After rambling through a few of the principal streets, and purchasing a few souvenirs of our visit, we again embark and float lazily down the lake. The famous old city of Constance soon appears in sight, and we stop long enough to admire its picturesque architecture, and magnificent cathedral, and to observe that the country around it is devoted largely to market gardening, producing such fruits and vegetables as are produced in Illinois.

The Aar River. From Lake Constance the river flows nearly west to Schaffhausen, where "it is precipitated over a ledge of rock, in three leaps, fifty or sixty feet in height," and then moves on calmly and quietly amid green woods, and in sight of many villages. We notice where the Aar river comes in, and wish we had time to ascend its waters, view its tributary lakes, and climb some of the mountains which look down upon it in solemn grandeur. We must be content, however, with recalling what the geographies say about it. From them we learn that the Aar rises near Mt. St. Gothard, flows northwest for about half its course, and is then turned to the northeast by the Jura mountains. From the west it receives the waters of Lake Neufchatel, on which is a city of the same name; and from the east come in the waters of lakes Zurich and Lucerne. These lakes, with many others, are visited every year by thousands of tourists who are charmed with the scenery of Switzerland. Even the barren and frowning mountains, with their dark defiles and forbidding chasms, are a source of revenue to the inhabitants. The income thus derived, added to the scant returns from their flocks and fields, from wood carving, and the manufactures of cotton, silk, tobacco, watches, and musical instruments, enables them to live in comparative comfort. Berne, the capital, is on the Aar, while Lucerne and Zurich are on lakes of the same name.

On returning to the Rhine, we continue our westerly course until we reach Basle, just as the river is about to quit the boundary between Germany and Switzerland, and turn to the north. The city is situated upon a terrace at the great elbow of the Rhine and is noted for its manufactures of cotton, silk, tobacco, chemical products, and ribbons. It is the leading commercial center of the country, its position with regard to France and Germany insuring it a large trade. Indications of peace and prosperity are visible on every hand.

The Middle Rhine. As we descend the river from Basle we are passing nearly through the center of a great valley once occupied, it is believed, by an inland lake. Looking to the west across Alsace, we see the Vosges mountains forming the rim of the valley in that direction; on the east the Black Forest mountains form the rim. The land near the river is low and flat, but farther back the surface is undulating, being influenced by ramifications from the mountains. We can see that it is carefully tilled. On the lowland near the river, wheat, oats, rye, barley, tobacco, hops, and the common fruits and vegetables are cultivated and yield abundantly, as the soil is enriched with manures and commercial fertilizers. The foothills are devoted to the culture of the vine, and so are the slopes of the mountains, wherever they can be terraced. We notice the general absence of fences, and on inquiring, learn that the land is too valuable to be wasted on fences or hedges, as boundary stones answer the purpose to the entire satisfaction of the people.

Many hamlets and villages are seen. They are occupied almost entirely by the tillers of the soil, and have no commerce or manufacturers. These industries are confined to the cities and towns. In some places the peasants are at work near enough to the river for us to notice their dress. Many of both sexes are barefooted; others wear wooden shoes and a very few, leather ones. Some of the men wear trousers, but many may be seen with knee breeches. The women wear a loose jacket or waist over very short skirts, and thick woolen stockings, provided they wear shoes. If they go without, they wear soleless stockings. The costumes usually change with the locality, each duchy or province having a costume peculiar to itself.

The first city of importance at which we arrive is Strasburg, the capital of Alsace, situated a short distance from the river on a small tributary. The industrial importance of this city is over-shadowed by its military renown. It is encircled by a network of fortifications, and each year adds to their strength. Our interest centers in the famous cathedral, with its tower and wonderful clock. Farther down, Mannheim appears at the junction of the Neckar river with the Rhine. It is a busy commercial hive, and standing, as it does, at the head of navigation for the larger class of river boats, its harbor is at all times crowded with vessels. On the Neckar, about twelve from Mannheim, is Heidelberg which claims to be the most beautiful town of all Germany. It is noted for its delightful surroundings, its old castle, and its university. Could we ascend the Neckar, we should find its source in the Black Forest mountains, and might be able to see the charcoal burners at their work.

Strasburg.

Descending the Rhine from Mannheim, we soon arrive at the old city of Worms, now back quite a distance from the river and noted in history as the place where Martin Luther appeared before the imperial diet to answer for his heresy. The Main river comes in on the right where the Rhine makes a sudden bend to the west. On it is Frankfort, about twenty miles from its mouth. This city is "one of the great money marts of Europe," being the headquarters of the Rothschilds. To scholars it is of interest chiefly as being the birthplace of Gœthe. Opposite the mouth of the Main is Mayence, strongly fortified and defending one of the most important passes over the Rhine. The bakeries in this city are said to be on a scale sufficient to supply the daily wants of 500,000 men. Its principal trade is

in wine, grain, and wood. As we pass by the city
we are reminded that here was born Gutenberg, the
inventor of printing, and that here dwelt the hard-
hearted Bishop Hatto who was punished in the
"Mouse Tower" for his cruelty. The island on which
stands the famous tower is near Bingen. In justice
to the memory of the bishop let us hasten to say for
aught that is known he was a good man, and was
not eaten up by rats and mice.

"I saw the blue Rhine sweep along,—I heard, or seemed to
  hear,
The German songs we used to sing in chorus sweet and clear:
And down the pleasant river, and up the slanting hill,
The echoing chorus sounded through the evening calm
  and still:
And her glad blue eyes were on me, as we passed with
  friendly talk,
Down many a path beloved of yore, and well-remem-
  bered walk!
And her little hand lay lightly, confidingly in mine,—
But we meet no more at Bingen,—loved Bingen on the
  Rhine."

Bingen. The village of Bingen stands on the left
bank of the river, and the hillsides back
of it are still covered with vineyards as in the day
when the dying soldier spoke of it so fondly. Here
the river turns to the north, and forces its way
through the mountains. From Bingen to Coblentz
the scenery is well-nigh beyond description. "The
rock-walls of the river; the continuous villages, the
quaint churches amid vineyards and cherry orchards, the mossy meadows about the mountains, the
white-kerchiefed villagers, present so many varied
and delightful objects, that the eye feasts on beauty,

and wonders expectantly at what the next turn of the river will reveal." The banks are lined with castles, villages, and ruins. In times past "every hill had its castle, and every crag its gray tower." Each old castle has its legends of robbery and rapine. River, shop, and farm were the prey of the robber-knights of the middle ages. Now all is changed. The knight is gone, the castle is in ruins, and German industry is protected by the strong arm of the law. The memories of knights and castles remain only as a source of inspiration to the poet and the story-teller.

The Lorelei. The most famous legend of the Rhine is that of the Lorelei, "the fairest of the fair," who sat on a bold promontory and by her magic music lured sailors to destruction in the rapids at the foot of the precipice. This interesting spot is on the right bank of the river and about midway between Bingen and Coblentz. As we pass by we listen for the echo of the cries of her victims, but it comes not, although it was said to be repeated fifteen times. Instead we hear the songs of the raftsmen as they bend to their long, sweeping oars; for enormous rafts form one of the sights of this classic river. We pass many of them on their way to the manufacturing districts. The logs are cut farther up the mountains, and floated down the mountain streams to the Rhine where they are fastened together. Each raft is a floating village. On it are erected rows of huts in which the two or three hundred persons on board eat and sleep. In addition, it often carries poultry, sheep, and a few cows, to furnish the people with a part of their food.

Coblentz. Coblentz is situated at the confluence of the Moselle and Rhine. It is surrounded by sterile, thinly peopled hills, and possesses few resources of wealth. On the opposite bank of the Rhine is the strong fortress of Ehrenbreitstein, which we are not permitted to enter. The military character of Coblentz overshadows and cripples its industrial development, and there is but little in the city that we care to see.

The sources of the Moselle are in the Vosges mountains, and considerable of its course is in France. On it is the strongly fortified city of Metz, which is now one of Germany's strong military out-posts.

From Coblentz the Rhine flows for about ten miles through a narrow plain which is semi-circled by mountains. It then enters a second defile which is less wild than that of Bingen. The Seven Mountains are on the right. Their sides slope gradually to the river, and are covered with vineyards, hopyards, and apple and cherry orchards. Drachenfels, ten miles southeast of Bonn, is the most famous of the Seven Mountains.

> "The castled crag of Drachenfels
> Frowns o'er the wide and winding Rhine,
> Whose breast of waters broadly swells
> Between the banks which bear the vine."

We have passed through the middle Rhine, the Rhine of legend and song, and soon reach Bonn on the outskirts of the great alluvial plain of Germany. It marks the northern limit of the vine, and is noted for its university and for the statue of Beethoven, who was a native of the place.

Cologne. Cologne is the next city of importance. It is the largest city of Rhenish Prussia, and has an extensive commerce both by river and rail. We pay but little attention to its manufacture of *eau de Cologne*, as probably more of the "genuine article" is manufactured in Chicago than in Cologne. But we visit its great cathedral and admire its beauty. We are shown the skulls of the three wise men of the East, who came to visit the Savior. The sexton is so sure that these are the wise Asiatic skulls, or rather the skulls of the wise Asiatics, that we express no doubt as to the truthfulness of his story. It is well that we do not, as a greater surprise is awaiting us. In Saint Ursula's church we are shown not only the skulls but also the rest of the bones of the 11,000 virgins who were slaughtered here by the Huns on their return from Rome. We pay the sexton his fee, feeling that he has earned it.

Leaving the presence of the dead we saunter back to the hotel and order dinner. It is served to us in a pleasant garden by the side of the river. Here we find a number of people eating, drinking, smoking, and chatting, each enjoying himself in the manner which best suits his taste. This decidedly German scene, supplemented by the swiftly flowing river with its steamers puffing hither and thither, and boats and barges passing to and fro, awakens pleasanter thoughts than did the bony fragments of wise men and virgins.

After dinner we travel by railroad to Aachen (Aix-la-Chapelle), which is nearly due west from Cologne, and close to the Belgian frontier.

Aix-la-Chapelle. The hot, sulphur springs of this place so pleased Charlemagne that he made it the capital of his empire. His marble palace has disappeared, but the chapel in which he worshiped, and in which he was buried still exists as a part of the cathedral. The rheumatic and the gouty visit the springs in large numbers annually, but the prosperity of the city is due mainly to its coal, lead, and zinc mines, and to its manufacture of woolen cloth, shawls, silks, leather, etc. Having visited the places of interest around the springs, we hasten back to the river, and again embark in our canoe.

The Lower Rhine. From Cologne to the North Sea, the Rhine passes through a low, level country, and the current becomes more and more sluggish. For the rest of its course in Germany, the scenery is uninteresting to the pleasure seeker; but to those who delight in the prosperity of the people, there is much to please. Well-tilled fields, and sleek herds, are seen in all directions, while at every bend in the river there is a thrifty town, which is a commercial center for the country back of it. Coal and iron are found in abundance, and, as a consequence, furnaces, rolling mills, cotton factories, silk factories, etc., are in nearly every town. Perhaps, the town in which governments are most interested is Essen, northeast of Dusseldorf. Here are manufactured the great Krupp guns. The guns, however, form but a small part of the products of Krupp's extensive establishment, which employs 20,000 men.

We now enter Holland, and the Rhine soon separates into several sluggish channels, which find their way between strong embankments to the North Sea. Following the usual custom of tourists, we spend no time in describing the Dutch Rhine, but proceed to Amsterdam, from which we take steamer to New York, and from there hasten to our homes.

CHAPTER XV.

A TRIP TO CEYLON AND INDIA.

Ceylon. On the morning of October 3, 18—, we came in sight of the island of Ceylon. The voyage from New York by way of the Suez Canal was a prosperous one. The first three days we spent in our state room, without any desire to see or to be seen. What took place there it is not necessary to relate farther than that Neptune was inexorable. On the afternoon of the fifth day we were able to go on deck and enjoy the congratulations of our friends, and by the eighth day we felt that we were able-bodied seamen.

We had read about Ceylon, its flora and fauna, its people with their customs and costumes, and were as we believed fairly well prepared to appreciate the beauty of the island,—and it is very beautiful. As we sailed south along the western coast, our expectations were more than realized. The south and west coasts are low and fringed with the cocoanut tree, which grows down on the water's edge. This tree is as valuable to the Cingalese as the reindeer is to the people of Lapland. It furnishes them with food, drink, clothing, and houses, and enters largely into their few simple arts and manufactures.

The east coast of the island is high and precipitous, and lacks the rich verdue of the south and west. It has no good harbor. Indeed there is no good natural harbor on the island. Point de Galle to which our ship is bound, is the principal stopping place for foreign vessels;* but its harbor is neither safe nor commodious. The British government is planning to build artificial harbors on both east and west coasts.

The anchor is dropped some distance from the shore and the Rosalind is soon surrounded by swarms of nondescript little crafts whose dusky occupants by shouts and gestures make known their willingness to serve us—for a consideration. The captain tells us that we shall have time to land and learn considerable about the island, while he is unloading a part of his cargo and taking on a large quantity of coffee and timber for Calcutta. So we enter one of the small boats and are soon safely landed. We hasten to the residence of the American consul, who kindly interests himself in our behalf, and provides us with six coolies who are to act as guides, and carry such supplies as we shall need. They soon appear, prepared for the journey, but to our disgust our official friend has hired six women instead of six men, as we directed. We objected to his choice, but he told us plainly that we must take these or go without any. After enjoying our look of dismay for a short time, he laughed heartily and made some remark about the emerald hue of Ireland, but we could not see that the vegetation of Ireland had any bearing on the case in hand. The consul at length assured us that these were all men, and there was not a woman among them. He frankly ad-

---

*Colombo on the west coast is now the capital and principal seaport

mitted that our mistake was a natural one, as the men and women are about of the same size, and dress alike. The principal garment, and often the only one, is a long, loose gown. And as both sexes wear their hair long, the men doing theirs up with combs, foreigners on their first visit usually mistake the men for the women.

Our visit to the interior, brief as it necessarily was, proved a profitable one. The island is quite mountainous in the south, while in the north it is level or undulating. Lying so near the equator, and being well watered, Ceylon is covered with a rich growth of vegetation. The cocoanut palm, already mentioned, is the most valuable of the trees, but large quantities of ebony, satinwood, and tamarind are exported annually. Plantations of nutmeg trees are carefully cultivated. The tree is about as large as a medium-sized apple tree. The cinnamon tree also flourishes here, as does the beech-like tree that bears the bread-fruit. The fruit does not grow among the branches, but out of the side of the trunk. Coffee is cultivated extensively, but rice is the principal agricultural product and the staple food of the people. Millet, sugar-cane, tea, pepper, pineapples, and tobacco are also produced and exported in considerable quantities

The natives live in villages which are over-shadowed by luxuriant woods. Their houses are constructed mainly of light bamboo poles, and roofed with the leaves of the palm. Each is surrounded by a small garden fenced in by palm leaves, or by hedges of flowers. Flowers are found growing in great profusion everywhere and at all seasons.

Wherever we go they greet us with their beauty and fragrance. In religion, nearly all of the people are Buddhists. Their temples are usually built on high, picturesque situations, and surrounded by pleasant grounds. The priests reside in the temples, where they instruct the youth in the mysteries of their religion. They are celibates and mendicants, and every morning some of them may be seen going to the village to receive offerings of rice, spices, etc.

A few precious stones are found in some districts, but none of any considerable value. We were told by our coolies that plumbago, quicksilver, and iron exist in great quantities. This may be so. Our desire to see as much of the island as possible kept us from investigating the geological formation very closely, and we saw no indications of any of these minerals. The pearl fisheries, mainly on the west coast, are very valuable.

We saw no large animals on our journey, but we were told that the elephant, bear, panther, and buffalo are found among the mountains. The buffalo is tamed by the people and used in plowing their land. Monkeys were seen frequently and in large troops. They were not glad to see us. They expressed their feelings in language more forcible than phonetic, and emphasized their remarks by dropping sticks and nuts from the treetops upon our heads.

All birds that we saw had very gay plumage, but we did not hear one sing in all the time we were on the island. Poisonous snakes are numerous, the principal one being the "anaconda," twenty or thirty feet long. Crocodiles inhabit the rivers and lakes, and are sometimes found in the artificial reservoirs in which are stored the surplus waters of the rainy seasons, of which there are two. The joyous, persistent mosquito impressed us most deeply of all the animals of Ceylon. Its affability and industry will not soon be forgotten.

The time which our captain had fixed for his departure from Point de Galle being near at hand, we returned to the coast. On the way we crossed several rivers and traveled along the margins of two lakes. Both lakes and rivers are small, and are not navigable for any but small boats. Bidding our friend, the consul, good-bye, we hastened on board our ship, and are now rounding the southern point of Ceylon, bound for Calcutta.

India. Our vessel stopped next at Pondicherry, to put off some wines and hardware, but we did not land. The city, with 112 square miles of territory, belongs to the French. It serves as a coaling station and as a depot of supplies for their ships in this quarter of the globe. At Madras there is a good artificial harbor constructed by the Indian government. We landed here and tried to see as much of the city as possible while the Rosalind was unloading some of her cargo.

We found little to interest us, although the vanity of the natives was somewhat amusing. They claim to be superior in many respects to the people of other parts of the country. And as others will not admit the claim, the inhabitants of this black, **Madras.** and not-overly-clean city seem to regard it a duty to impress strangers with a sense of their superiority. We were duly impressed.

and returned to the harbor where for about an hour we watched the native boys sailing their boats, or, perhaps more properly speaking, their rafts. Each raft consists of three logs about ten feet long, and tied together with cocoanut strings. We expected every moment to see some of them dashed to pieces by the heavy surf that prevailed outside the storm wall, but we were happily disappointed. The same good angels who watch over the destinies of boys in America and Europe were evidently caring for the "brownies" of Madras. The principal exports are coffee, cotton, sugar, grain, indigo, and other dyes. Nearly all kinds of European manufactures are imported.

The Hoogly River. After a pleasant voyage along the southeastern coast of India, with the Eastern Ghauts in sight most of the time, we entered the Hoogly river. This is the western mouth of the Ganges and the most direct water-way to Calcutta. East of it is the great delta formed by the Ganges and Brahmaputra rivers. It is cut up into many islands by innumerable channels. The islands bordering on the bay of Bengal are known collectively as the Sunderbunds. Rice is the principal crop and the staple food of the people. Several of the islands are wooded, and the woods are infested with serpents and tigers which often cause great destruction of property and of human life.

Calcutta. We were fortunate in reaching Calcutta in the day time, and in not being troubled by the "bore" which sometimes causes great destruction on the Hoogly. The city is built on low land, consequently the natural drainage is poor. Notwithstanding its low and malarial position, however, engineering skill has succeeded in rendering it fairly healthy.

On landing, we inquire the way to the American consulate. A native policeman kindly guided us to the door. Our passport served as a letter of introduction, and on reading it the consul generously offered his services to make our stay as pleasant as possible. In the afternoon he took us in his carriage to see the most noted sights. The drive to the beautiful suburb of Garden Reach was very enjoyable. The elegant country seats and picturesque gardens with their wealth of tropical plants and flowers, would have to be described in the glowing language of the Orient to do them justice. We met and Indian Rajah, or native prince, with his retinue. He was mounted on a richly caparisoned English hunter, and made quite an imposing figure with his flowing silken robes and jeweled turban.

The Brahmins. We met many Brahmins on their way to the temples. Their features were as clear-cut and regular as those of any European or American. They were not black, or even brown, but slightly bronzed. With bare heads cleanly shaved, and with arms and breast naked, they walked with self-conscious pride as if they would impress all who saw them with the fact that they were members of the most ancient aristocracy upon earth. The simple string that hangs over a Brahmin's shoulder and across his breast confers a higher honor than does the insignia of the Garter, or of the Golden Fleece.

But all of the natives are not Brahmins; many belong to the lower castes. These are darker, many being almost black, and all have that timid, cringing look peculiar to the oppressed of all lands.

There are two Calcuttas. One is the native city with narrow unpaved streets and with bamboo or mud houses, low and filthy and swarming with naked or semi-naked humanity. As we were driving through the streets, the people were preparing their evening meal. The fuel used by the poor is cow-dung, and the smoke which it gives out is exceedingly unpleasant, consequently we were glad to enter the English Calcutta. This is well-built, with wide streets and spacious avenues. The houses are large; many of them of brick, covered with stucco, and having broad verandas.

Calcutta is not only the chief seat of government for India, but also claims to be the greatest commercial city of Asia. It receives the products brought from the interior by the Brahmaputra and Ganges, and from the Sunderbunds by several canals which pass round and through them. Besides it has three trunk lines of railroads, each having several branches or feeders. Its chief exports are jute, opium, indigo, rice, wheat, hides, cotton, raw silk, and tea. The imports are cotton goods, linens, hardware, pig-iron, silver, wine, and salt. Population, about 1,000,000.

The Ganges River. When we started from Calcutta for northern India, we could have traveled by railroad as there is a well-equipped line connecting the capital with the cities of the north. It runs quite close to the Ganges, passing through Benares, Allahabad, Lucknow, and Delhi; and then curving well to the northwest, it ends at the Chenaub river, one of the principal tributaries of the Indus. As our purpose was, not to travel through the country simply, but to see as much as possible of the people and their manner of life, we decided to travel up the Ganges in one of the native boats. The increasing shoals in the river, and the cheap communication by railroad have led to the giving up of steamboat navigation above Calcutta. The native boats carry on an immense traffic. It is estimated that 60,000 crafts of all shapes and sizes pass up or down by Benares every year. Our boat had a thatch roof over part of it, to protect us from the heat of the sun. This thatched part was our kitchen, dining-room, and parlor by day, and our bedroom by night. It was not palatial, but we were not proud. The view of the adjacent country was very satisfactory, as the banks of the river, for hundreds of miles, are low. Our crew could talk English quite readily, and were willing to give us all the information they could. When the wind was favorable, an old ragged sail was used, but when unfavorable, poles or "sweeps" were used to propel the boat, and many times the crew walked on the shore and pulled the boat by means of ropes. The downward passage is made more easily. We rarely traveled at night. The boat was usually tied to the bank at dusk and started again at daybreak.

Benares. The first city of note at which we stopped was Benares, the holy city of the Hindoos. We staid here two days, and saw much that was of interest. The city is to the Hindoos what Mecca is to the Mohammedans. Devout people from all parts of India come to worship at its shrines, and many come to die. When a pious Hindoo feels that his end is near, he has his friends convey him to the city to die, for then his spirit will pass at once to Brahma. If he dies, they burn his body on the funeral pile and scatter his ashes on the Ganges. If they do not remain by until the body is burned, the probabilities are that the unscrupulous attendants will throw it into the river in a semi-charred state, in order to save the fuel for the next subject. Many such bodies floated by us on our way up here, but we did not understand their significance until we saw the scenes that were being enacted in front of Benares. Incredible as it may seem, sons have been known to choke their dying father with mud from the sacred Ganges, or to break open his skull so as to let the spirit out, and then to push the body into the river while it was but slightly charred.

The largest and holiest temples are near the river, and imposing flights of marble stairs lead down to the water. Here may be seen at all hours of the day hundreds of devotees bathing, or standing up to their armpits in the water, calling upon Brahma and repeating passages from their sacred books. We visited what may be termed the leading theological school, the most famous bazaar, and the temple dedicated to the monkey god. In our rambles we passed through scenes of splendor and squalor, wealth and poverty, Paradise and Gehenna, side by side. In no other quarter of the globe are extremes found in such close juxtaposition as in

Asia. This is one of the distinguishing characteristics of the continent.

As we ascend the river above Benares, we have an excellent opportunity to see the country. The boat moves so slowly that when it happens to be rounding a bend in the cool of the evening, we frequently land and walk across, and sometimes travel several miles inland. Nowhere do we see any isolated farm houses. The land being cut up into little farms of two or three acres each, makes it possible for the people to live in villages. The tillers of the soil, however, are not the only inhabitants of the villages. In many of them may be found weavers, leather-dressers, shoemakers, and others engaged in what may well be termed unskilled labor. The houses are usually built of mud, or unburned brick, and the roof of baked clay. They have no doors or windows facing the street, as that would make it possible for strangers to see their women, a degradation that is to be feared more than death.

Character of The Farms.

The people of India are very poor. There are a few wealthy nobles and grandees, and a comparatively small number who are in comfortable circumstances, but the great mass of the people are poor, very poor. It is doubtful if half of the peasants of India ever eat enough to satisfy their hunger. Two slight meals a day seems to be the regular allowance for most of them. Scarcity is the most constant visitor at their homes. It cannot well be otherwise in the home of the day laborer who works for six cents a day, and has to support a family of five or six persons on such

Poverty of The People.

a pittance. Were it not that the climate is such that but little clothing is needed, and only fuel enough to cook what little food they have, the poor could not keep body and soul together.

An Illinois farmer, driving for the first time through the agricultural regions of India, might well be astonished at the patchwork appearance of the crops. Here an acre or two of wheat; adjoining it two or three acres of millet; then a patch of cotton, and next a small field of indigo, or Indian corn, with an occasional strip of barley. These are the principal crops, and they are often interspersed with diminutive areas of potatoes, cabbages, and onions. When we consider that there are no fences separating these fields, we must not be surprised if our Illinois friend regards the patchwork as decidedly on the "crazy-quilt" order.

The culture of coffee is carried on extensively in some parts of the country, but it is mainly in the hands of foreigners. The same is true of tea, whose culture is confined to the valley of Assam. The sugar produced is not sufficient for home consumption. Many of my readers may suppose that rice is the chief crop. This is a mistake. It is cultivated only in the deltas of the great rivers, on the strip of lowland along the western coast, and over a small area in the northwest. It is not the staple food of the people of India as a whole. Millet is their main dependence. The opium poppy is cultivated in the vicinity of Benares and Patna, and on the northern slopes of the Vindhya mountains.

The Mela. When we arrived at Allahabad, the great Mela was in session. These Melas were originally religious convocations, a sort of Hindoo campmeetings. They have been partly secularized, and are now combinations of religious meetings and fairs. A temporary city of booths and tents was erected on the tongue of land between the Ganges and the Jumna rivers. It was estimated that over half a million people were in attendance.

It was a motley crowd, embracing all classes and conditions, although the majority were of the common people. Fakirs (religious vagabonds), merchants, and jugglers plied their several vocations. Thousands of men standing in the water were busily engaged in religious exercises. Husbands could be seen leading their wives with covered heads into the river, where they, too, engaged in the rites of their religion. All of these we respected, for evidently they were sincere. But the Fakirs! How we longed for a strong force of lusty, physical missionaries, with a plentiful supply of scissors, combs, and soap, and authority to use them. The Fakirs were the filthiest animals we saw in India, or elsewhere. And the filthier they were, the greater was their odor of sanctity, in the estimation of the people. On the whole the sights witnessed at this great gathering were pitiful, saddening, and disgusting, and we willingly paid our boatmen an extra fee to induce them to leave a few days before the time set for our departure.

The Sepoy Rebellion. From Allahabad to Cawnpore the voyage was uneventful. Here we bade goodbye to our faithful crew, and with becoming decorum received their salaams. After refreshing ourselves at a neat English hotel, we hired a guide to show us the scene of the Cawnpore massacre. As we looked down into the fatal ravine that opens out on the Ganges, we could, in imagination, see the small body of brave British soldiers, with the women and children in the center, marching down to the river. They had for several days resisted successfully the archfiend, Nana Sahib, and his Sepoys. Bravely they fought although their number was small and their strength gone because of exposure and lack of food. No man thought of surrendering, for all knew that not only the lives, but what was more precious, the honor of British women, was at stake. At length the Nana seeing that it would take more time to overcome them than he could spare, determined to destroy them through treachery. So he offered to let them go down the river to their friends and to furnish boats for the women, children, and wounded. The offer was in writing over his signature, and unfortunately was accepted. As they approached the ravine they could see the boats in readiness, with the boatmen standing by. They had no sooner entered the ravine, however, than a blast from a bugle was heard in the rear, and instantly masked batteries opened upon them. The attendants set fire to the boats, and five hundred rifles poured their missiles upon the devoted troops. The carnage was over in a few minutes. Only four of the soldiers escaped to tell the story. The Sepoys were instructed to spare the lives of the women, and they and the children were imprisoned in a small building of two rooms. On hearing that General Havelock was coming to their rescue, Nana ordered his troops to kill all of them. This they refused to do, so he hired five butchers from the town to do the deed. These monsters entered with swords and knives, fastened the doors behind them, and in an hour and a half their work was finished. The common scavengers dragged out these once beautiful women and children and threw them into an open well near by. We visited the well. It is still the grave of those so cruelly massacred; and over it the great British nation has erected an elegant building which is surrounded by most charming grounds.

CHAPTER XVI.

A TRIP TO CEYLON AND INDIA.—CONTINUED.

Lucknow. From Cawnpore we traveled overland to Lucknow, the scene of the famous siege. As we stood within the Residency it required no great effort of the imagination to people it once more with brave soldiers, Christian women and helpless children. We could hear the commander, Sir Henry Lawrence, ordering the men with his dying voice, to "save the ladies," and never to surrender. And we could hear them, in turn, pledging themselves that there should not be another Cawnpore, while wives were heard exacting promises from their husbands that when death became inevitable they would kill them with their own hands rather than let them fall alive into the hands of the Sepoys. We asked about the truthfulness of the story of Jessie Brown, and our guide, a one-legged veteran, and evidently a Highlander, declared that the story is true, no matter what historians may say to the contrary. As we paid for this information, it is but reasonable that we should use it.

> "There Jessie Brown stood listening
> Till a sudden gladness broke
> All over her face; and she caught my hand
> And drew me near and spoke:—
>
> "'The Hielanders! O, dinna ye hear
> The slogan far awa?
> The McGregors',—O, I ken it weel;
> It's the grandest o' them a'!

> " 'God bless the bonny Hielanders!
> We're saved! we're saved!' she cried,
> And fell on her knees; and thanks to God
> Flowed forth like a full flood-tide."

The Taj Mahal. To leave India without visiting Agra and inspecting the Taj Mahal would be sure to bring upon us the censure of our æsthetic friends. We therefore took the train from Lucknow and soon arrived in sight of the most beautiful building in the world. To attempt anything like an adequate description of the Taj would be a helpless task. Artists and critics who have seen it, acknowledge that language is too feeble to describe its matchless beauty. One enthusiastic traveler says that, "Viewing it from the lofty tomb of Akbar, five miles distant, it looks like a tent of snowy whiteness and rich embroidery let down from heaven into a paradise of earth to be the audience-chamber of an angel on an errand of mercy to men." Others are quite as enthusiastic, though perhaps not so poetic. We had read several descriptions of the building, and supposed we were ready to appreciate its exquisite grace and symmetry. But as we stood in its presence, we felt that our ideas of the structure fell far short of the reality. We forgot the descriptions and stood entranced by the divine influence of the builder's art. Several times did we try to turn away from the Taj to admire the well-kept grounds with their rows of cypress trees, orange, lemon, and palm, and the profusion of choicest flowers which border the walks, but in vain. Our eyes were not satisfied to dwell on the inferior while the superior was present; and in this instance man had excelled nature.

Delhi. The journey from Agra to Delhi is a short one, and is made over a good railroad and in a comfortable car. Delhi was long the center of Mohammedan power in India, the capital of the Great Moguls. Here was erected the Peacock Throne at a cost of $150,000,000, but it was carried away long ago by the Persian invader. The Sepoy rebellion put an end to the rule of the Great Moguls. Many of their palaces are in ruins, while those that remain are occupied by British officials. On the streets we saw many snake charmers and jugglers trying to obtain a few coins by entertaining the populace. Here, also, through the kindness of a British captain, who liked America and Americans, we met several intelligent Hindoos and Mohammedans, from whom we obtained much valuable information in regard to the country, and the hopes and aspirations of their people.

The Women of India. Our readers will understand, no doubt, why there is so little said in these notes about the women of India. Custom excludes them from the sight of all men except those of their own immediate family. The "Nautch girls," that is the dancing girls, may be seen in any of the large cities. They are social out-casts, yet the most intelligent class of native women. At feasts and merry-makings they are hired to entertain the guests with their witticisms, singing, and dancing. But even they cannot be hired to engage in a dance with persons of the opposite sex. Much less could any respectable woman be induced to do so. How Americans and Europeans can permit their wives and daughters to take part in mixed dances is one of the mysteries which the eastern mind cannot fathom.

The Vale of Cashmere. Bidding farewell to our friends at Delhi we took the train to Moultan, on the Chenaub river, one of the principal tributaries of the Indus. Here we hesitated some time as to what course to pursue next. We desired very much to go north to Cashmere.

> "Who has not heard of the Vale of Cashmere,
> With its roses the brightest that earth ever gave,
> Its temples, and grottos, and fountains as clear
> As the love-lighted eyes that hang over their wave."

We had heard of the famous vale, had read about the skill of its weavers and silversmiths, and would have liked to see for ourselves the products of their skill but the brief time at our command forbade our doing so. Extravagant as Moore's description may seem, modern travelers, who have visited the region, tell us that the poet's fancy did not far exceed the reality; that it is a veritable paradise surrounded by walls of snow-capped mountains, a smile of heaven set in the midst of nature's frowns. They also tell us that the valley is not only famous for its scenery, but for the productiveness of its soil as well. Wheat, maize, barley, and millet are the principal crops, but the common fruits and vegetables are cultivated with profit.

The Indus River. At Moultan we embarked on the Chenaub river, and as we were borne along by the current, we recalled the fact that more than twenty-two hundred years ago Alexander the Great sailed over the same course. It was at the site of Moultan that he came near losing his life in trying to take the stronghold of the Malli. And it was at the junction of the Chenaub with the Indus that he ordered extensive dockyards to be constructed and his last Alexandria to be built. The volume of the Indus grows less and less as we approach the delta, owing to its passing through a rainless region, the eastern outskirts of the desert of Gedrosia, which proved so destructive to the Macedonian hosts. Notwithstanding its discouraging natural conditions, much of the land on both sides of the river, back as far as we could see, is rendered productive by means of irrigating canals. The delta is very fertile, another Egypt, and like Egypt is fertilized every year by the river which made it.

We left the Indus at Hydrabad and traveled by rail to the great city of Bombay. Everywhere along the route, we could see the farmers trying to coax or force a scanty subsistence from the earth. In many places we could see carts drawn by a single ox, carrying last year's crop to market, or conveying the family on a visiting trip. When the occupants were women of the higher or middle castes, the cart had a canopy over it as a protection from the sun, and curtains at the sides as a protection from the gaze of the unclean, by which is meant persons of the lower castes and foreigners, whose glance is to be dreaded more than the fiercest rays of the sun.

Bombay. Bombay, the second city of India in size and importance, is situated on the western coast of the peninsula. It is the great commercial rival of Calcutta, but is handicapped by the fact that its connection with the interior is entirely by railway, while its rival not only has an extensive system of railroads, but is also able to reach the most fertile parts of the country by means of the Brahmaputra and Ganges rivers, with their tributaries. Bombay is still further hindered in the commercial race by the fewness of its exports; cotton, grain, and opium being the only articles worthy of mention, while Calcutta has a great variety. Its nearness to the Suez canal, however, together with its good harbor, offsets in part the advantages of the city on the Hoogly.

When the train arrived at the well-appointed union station, we could easily imagine ourselves in one of the large cities of America. The offers of cabs, carriages, and "buses" were as many and as hearty as we ever had made to us in Chicago, and they were made in fully as many varieties of English. Engaging a carriage and claiming our "luggage," we drove over smoothly paved streets to an excellent hotel. Next morning we started out shortly after breakfast to view the city. It is the most English of all the cities of India that we visited. The houses of the Europeans and the wealthy natives are large and commodious, and as a general thing each is surrounded by beautiful grounds, or "gardens," as they are termed here. The stores on the leading business streets are three or four stories high, and well stocked with the products of both the Orient and the Occident. And although there is much filth and poverty in the native quarter of the city, yet the conditions are not so dreadfully disgusting as in Calcutta.

The most pleasing sight that met our gaze was an American flag floating from the masthead of a steamer that was unloading an assorted cargo of machinery, farming implements, and railroad supplies. We had seen the flag waving its protecting folds over the consular residences in all the great commercial centers that we visited; but it meant more to us in this instance since it was recently from home, and would soon return again. The sight suggested a slight touch of home-sickness, and we determined to finish our itinerary as quickly as possible. We felt, however, that we must visit the Deccan and see for ourselves something of the ravages made by the famine and the bubonic plague.

The Famine. A day's travel by railroad brought us to Hyderabad in the south-central part of the peninsula. The sights that we saw were heart-rending; men, women, and children dying for the want of something to eat, while the world had more than food enough for all its inhabitants. We thought of the corn cribs of Illinois, full to overflowing with nourishing food, and enough of it being wasted by rats to save millions of human lives in this poor miserable country. It is sad, it is pitiful, that strong men, helpless women, and innocent children must die of starvation and rot by the roadside, when a little of the world's superabundance would save their lives. The government of India is doing much to relieve the suffering, and so are the the missionaries, but notwithstanding their efforts the condition of the poor is terrible. And this is not due to their want of thrift as many may suppose, but to the lack of sufficient rain. The Deccan peninsula being bordered by the eastern and western Ghauts, it often happens that not enough of the rains, brought by the monsoons, is permitted to reach the interior plateau to render the soil productive. When such conditions prevail, they are sure to be followed by famine, and famine is always the fore-runner of a plague of some kind. It is so now, and the twin sisters are gloating over their work.

Child Marriage. On our return to Bombay we engaged passage on the steamer Brunswick bound for Liverpool. As it would not leave for two or three days we determined to use the time in writing some general notes on India. Before doing so, however, we wish to say that we did not witness a marriage or a funeral service while in the country. We understood that child marriage still prevails; that it is no uncommon thing for girls of eight or ten years to be married to boys of like age, or even to men of twenty, thirty, or forty years; and that the parents would consider themselves disgraced if their daughter was not married, or at least betrothed, by the time she was twelve years of age.

Disposal of the Dead. In regard to the manner of disposing of the dead, we have already stated that the Hindoos burn theirs. For many ages it was the custom to burn the living wives on the funeral pile of their dead husbands; but this practice of sutteeism has been suppressed by the British. The Mohammedans always bury their dead: the poor in shallow graves, many of which are robbed by the jackals that infest the country: the rich in costly mausoleums. The Parsees of Bombay have erected a high tower, across whose top is a heavy grating.

On this grating they place their dead, and the vultures gorge themselves on the flesh. This seems the most shocking custom of all; and yet the Parsees are among the most intelligent and enterprising portion of the community.

Dimensions of India. India, the middle one of three great peninsulas that project south from Asia, has natural boundaries throughout. On the north are the Himalaya Mountains, on the west the Suliman separate it from Afghanistan and Beloochistan, and on the east it is separated from Burmah by spurs of the Himalayas; while on the southeast and southwest it is washed by the Bay of Bengal and the Arabian Sea, respectively. From north to south it is 1,900 miles, or about five times the length of Illinois. Its area is about twenty-seven times as great as that of Illinois, and its population more than three times that of the entire United States. This vast empire is governed by Great Britain; for although there are a few so-called independent states governed by native rulers, yet even they are subject to British control.

Surface. The surface is divided into three distinctly marked divisions. In the north, bordering on the Himalayas, is a belt of low mountains and foothills with an average width of a hundred miles. This region is occupied by various tribes of mountaineers; and the British officials and others who can afford to do so, spend several weeks here each year to recuperate from the depressing influence of the lowlands. South of this hilly tract is a great plain thirteen or fourteen hundred miles long and from two hundred to three hundred miles wide, and as level as a house floor. This plain consists of alluvial deposit washed down from the Himalayas and from Thibet, and is the most fertile and best cultivated part of the country. It is the home of the Hindoos, and was wrested by their Aryan ancestors from its earlier occupants.

Farther to the south, and separated from the great central plain by the Vindhya mountains, is the peninsula of Deccan. It consists of a great plateau bounded on the east and west by the Ghauts mountains, which unite at Cape Comorin. The inhabitants belong to the Dravidian family, and are supposed to have been driven south by the conquering Aryans.

Rivers. Of the great rivers of India, the Brahmaputra and Indus have their origin and much of their course north of the Himalayas, but they bestow their richest gifts on India. T e Ganges, the sacred river, rises on the south slope of the Himalayas, at the foot of a great snow-bed, 10,000 feet above the level of the sea. It is 1,500 miles long, and its average width is greater than that of the Mississippi. With all due respect to the deity whose name it bears, it is a treacherous stream, full of shoals and shifting mud-banks, and not fit for ordinary navigation.

Climate. The climate is hot. At no time of the year is it safe for Europeans to be exposed to the sun bareheaded. But while all seasons are hot, the greatest heat prevails from the middle of May to the middle of June. This is a trying time for foreigners, and were it not for the pankas, or fans, with which the houses are fitted up, the heat would be unendurable. There is no use for fires, except for cooking purposes, therefore the houses are built without chimneys, as the cooking is done in detached kitchens, or sheds.

In the rainy seasons the country is covered with verdure, but as the cultivated crops have already been discussed, it only remains to be said that great care is taken of the forests. They are in charge of skillfully trained foresters, who are government officials. These functionaries not only care for the forests, but it is also a part of their duty to see that trees are planted along the country roads. As a consequence, one may drive for hundreds of miles between rows of mango trees, valuable both for their shade and fruit. The jungles are numerous; but it should be borne in mind that "a jungle in India means any portion of wild land, whether covered with grass, bush, or timber." The jungles are the homes of many wild animals. The elephant, lion, tiger, leopard, hyena, jackal. etc., abound, and often commit great depredations upon the growing crops; while some of them even love to feast upon the owners. Serpents also are numerous, and disposed to court human society. The poorer natives, going barefooted and barelegged, suffer most from their venemous bites; foreigners are rarely injured.

Minerals. Our early reading led us to believe that India was a very rich country. In imagination we could see the mines of Golconda glittering with diamonds, while gold was so plenty that it was scarcely worth digging. The fact is, it is a comparatively poor country. Its soil is productive in the plains, but on the plateaus it is thin and poor.

and a good crop is the exception. There are but few places where it pays to mine for gold and silver. Iron is very plentiful, but because of the want of coal it is cheaper to import the metal from Europe than it is to attempt to mine and smelt the native ore. Coal has recently been found in a few localities, and if it should prove of good quality, it would be of more value to India than all its gems and precious metals.

Manufactures. The manufacturers are few and feeble. Some years ago European capital and skill attempted the manufacture of cotton-goods. The enterprise was very successful, so much so that the cotton kings of Manchester were alarmed. They raised a great hue and cry about the cheap labor of Asiatics being brought into competition with English labor. Their cry was listened to, and matters were so adjusted by the home government that the factories of India had to close their doors, and the poor of the country had to buy their few cotton garments at Manchester prices.

But the whistle of the Brunswick is blowing, and that means that passengers must be on board in an hour. So we close these notes, pack our grips, and say good-by to our readers.

www.ingramcontent.com/pod-product-compliance
Lightning Source LLC
Chambersburg PA
CBHW020307170426
43202CB00008B/530